Extinction: A Very Short Introduction

T0130767

VERY SHORT INTRODUCTIONS are for anyone wanting a stimulating and accessible way into a new subject. They are written by experts, and have been translated into more than 45 different languages.

The series began in 1995, and now covers a wide variety of topics in every discipline. The VSI library currently contains over 600 volumes—a Very Short Introduction to everything from Psychology and Philosophy of Science to American History and Relativity—and continues to grow in every subject area.

Very Short Introductions available now:

ABOLITIONISM Richard S. Newman
ACCOUNTING Christopher Nobes
ADAM SMITH Christopher J. Berry
ADOLESCENCE Peter K. Smith
ADVERTISING Winston Fletcher
AFRICAN AMERICAN RELIGION
    Eddie S. Glaude Jr
AFRICAN HISTORY John Parker
    and Richard Rathbone
AFRICAN POLITICS Ian Taylor
AFRICAN RELIGIONS
    Jacob K. Olupona
AGEING Nancy A. Pachana
AGNOSTICISM Robin Le Poidevin
AGRICULTURE Paul Brassley
    and Richard Soffe
ALEXANDER THE GREAT
    Hugh Bowden
ALGEBRA Peter M. Higgins
AMERICAN CULTURAL HISTORY
    Eric Avila
AMERICAN FOREIGN RELATIONS
    Andrew Preston
AMERICAN HISTORY Paul S. Boyer
AMERICAN IMMIGRATION
    David A. Gerber
AMERICAN LEGAL HISTORY
    G. Edward White
AMERICAN NAVAL HISTORY
    Craig L. Symonds
AMERICAN POLITICAL HISTORY
    Donald Critchlow
AMERICAN POLITICAL PARTIES
    AND ELECTIONS L. Sandy Maisel

AMERICAN POLITICS
    Richard M. Valelly
THE AMERICAN PRESIDENCY
    Charles O. Jones
THE AMERICAN REVOLUTION
    Robert J. Allison
AMERICAN SLAVERY
    Heather Andrea Williams
THE AMERICAN WEST Stephen Aron
AMERICAN WOMEN'S HISTORY
    Susan Ware
ANAESTHESIA Aidan O'Donnell
ANALYTIC PHILOSOPHY
    Michael Beaney
ANARCHISM Colin Ward
ANCIENT ASSYRIA Karen Radner
ANCIENT EGYPT Ian Shaw
ANCIENT EGYPTIAN ART AND
    ARCHITECTURE Christina Riggs
ANCIENT GREECE Paul Cartledge
THE ANCIENT NEAR EAST
    Amanda H. Podany
ANCIENT PHILOSOPHY Julia Annas
ANCIENT WARFARE
    Harry Sidebottom
ANGELS David Albert Jones
ANGLICANISM Mark Chapman
THE ANGLO-SAXON AGE John Blair
ANIMAL BEHAVIOUR
    Tristram D. Wyatt
THE ANIMAL KINGDOM
    Peter Holland
ANIMAL RIGHTS David DeGrazia
THE ANTARCTIC Klaus Dodds

Available soon:

For more information visit our website

www.oup.com/vsi/

Paul B. Wignall

EXTINCTION

A Very Short Introduction

OXFORD
UNIVERSITY PRESS

OXFORD
UNIVERSITY PRESS

Great Clarendon Street, Oxford, OX2 6DP,
United Kingdom

Oxford University Press is a department of the University of Oxford.
It furthers the University's objective of excellence in research, scholarship,
and education by publishing worldwide. Oxford is a registered trade mark of
Oxford University Press in the UK and in certain other countries

First edition published in 2019

Impression: 1

Published in the United States of America by Oxford University Press
198 Madison Avenue, New York, NY 10016, United States of America

British Library Cataloguing in Publication Data
Data available

Library of Congress Control Number: 2019939567

ISBN 978-0-19-880728-5

Printed in Great Britain by
Ashford Colour Press Ltd, Gosport, Hampshire

Contents

List of illustrations

List of Table

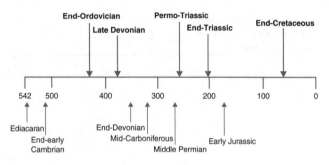

Time line of extinctions with mass extinctions above and lesser crises below

Chapter 1
Why extinctions happen

The idea of extinction

Most of us are familiar with the idea that species can go extinct.
Everyone has heard of the death of the dodo, and the violent end
of the dinosaurs provides the iconic example of extinction on a
vast scale. In fact extinction is commonplace, it is estimated that
more than 99.9 per cent of all the species that have ever lived
are now extinct. But the main concern today is that extinction
rates are accelerating way beyond the normal, background
rates to the point where we may be living at the start of a great
human-driven mass extinction event. As a consequence,
current research on extinction and its causes is at the heart of
conservation studies and similarly much effort is invested in
looking at why extinctions happened in the past, particularly
during the spectacular mass extinctions. This book looks at what
we currently know about causes and styles of extinction today and
in the past. This knowledge lies at the heart of our understanding
about the way the Earth's environment works and what happens
at times of extremes.

Extinction worries have not always been with us. Only 200 years
ago the notion that species could disappear was regarded as
unlikely—species were fixed, they did not evolve and they did not
die. This was despite the fact that fossil hunting, a trendy new

pastime, was turning up many new forms that no one had seen alive. Rather than assume their extinction, these discoveries were being added to a rapidly lengthening list of undiscovered species that were presumed to be living in remote, unexplored parts of the world. George Cuvier, French naturalist and founding father of palaeontology, challenged this orthodoxy. In his 1813 work *Essay on the Theory of the Earth* he proposed that the Earth had experienced a series of catastrophes (such as biblical-scale deluges) that had wiped out the incumbent species and replaced them with a whole new set of species. Cuvier's ideas were challenged by many, especially the British geologist Charles Lyell, who favoured much slower, more gradual changes in Earth history. A rich and long debate ensued, but by the mid-19th century Lyellian gradualism held sway and extinction, especially the catastrophic episodes favoured by Cuvier, was no longer on the scientific agenda. This, though, was about to change.

Darwin's publication of *On the Origin of Species* in 1859 is justifiably renowned as the birth of evolutionary theory, but less well-known is the fact that it also provided a mechanism for extinction. Evolution by natural selection occurs by the accumulation of slight improvements which gives the evolving species a competitive advantage over existing, unchanged species, particularly those that are closely related and therefore likely to have a similar lifestyle. Thus, for Darwin the appearance of a new species was seen as bad news for existing species and, as he put it, 'extinction of less-favoured forms almost inevitably follows'. This reasoning stems from Darwin's viewpoint that global diversity has been more-or-less constant through geological time, because all habitats have an essentially full quota of species, with the result that adding new species to the total inventory is a 'one-in one-out' process. Over 150 years later we inevitably have a more nuanced view of how global diversity has changed through time, which shows that there have been periods of rapid increases of species numbers, especially after mass extinctions, but conversely there are also long intervals of little change, as Darwin originally envisaged.

In many ways Darwin, a Victorian gentleman, was a man of his time who saw long-term progress as both the product of human endeavour and nature's selection process. Thus, he did not think extinction could be caused by environmental changes because such physically driven losses would likely lead to rather random evolutionary trends instead of the long-term improvement of life that he saw. Also, as an adherent of Lyell's gradualist school, he thought it unlikely that any abrupt, Cuvier-style mass extinctions had happened, although he acknowledged that, in special circumstances, multiple extinctions may occur. For example, the sudden immigration of many species into an area could cause losses among the local species. Overall though, Darwin's view of evolution and extinction was one of inexorable improvement to an essentially constant total pool of species that saw the gradual demise of 'inferiors in the struggle for existence' in favour of species that were 'higher in the scale of nature'.

Fast forward to today and our ideas about the causes of extinction have changed substantially (while Darwin's theory of evolution is of course well-established and has stood the test of time). Darwin's dismissal of catastrophic extinction events was still current up to 1980. Such mega-crises are now well-documented and they have given life's history a much more random appearance than Darwin ever envisaged. We now know that successful, dominant groups, such as the dinosaurs, can disappear and be replaced by obscure animals such as the mammals. Darwin's interspecific competition is still considered a valid process of extinction but habitat loss is now thought to be much more important.

Species–area relationship

In a celebrated series of experiments in the 1960s Robert Macarthur and E.O. Wilson showed that the number of species found on a series of small islands was related to the area by the following relationship:

$$S = cA^z$$

where S is the number of species, c is constant, A is the habitat area where the species live, and z is the slope of the curve when plotted on a log-log scale. This quantified a long-recognized fact that larger areas tend to have more species present than smaller areas, but it is not as simple as saying double the area, double the species. As the area increases, the rate of addition of new species flattens off, giving a species–area curve a convex-up appearance. In the natural world this power law relationship, as it is known, makes sense because once an area has attained a large size it is unlikely that there will be many new habitat types to appear for new types of species to occupy. Plotting a species–area curve using logarithmic axes produces a straight line with a constant slope (z) which is easily measured.

Macarthur and Wilson provided an explanation for the species–area effect. They suggested that the number of species present on an island represents a balance between new species arriving (immigrating), either by flying there or as seeds dispersing in the wind, and local species going extinct. The immigration rate is dependent on how far the island is from other islands and the mainland, both of which provide the pool of new species. The rate of extinction is dependent on an island's area, which controls the population size—you can only squeeze so many individuals onto an island. Larger islands can also accommodate more species because they are likely to have a greater diversity of habitats. The smallest islands may just have a few palm trees—as depicted in cartoons of shipwrecked sailors—while large islands may have volcanic uplands as well as coastal forests. The species–area relationship (SAR) is widely used by biologists and conservationists because it allows species numbers to be predicted from the area of habitat without the (expensive and time-consuming) need to actually visit and count all the species present.

The idea that extinction risk is related to habitat area has parallels in Darwin's 'one-in one-out' view, although it differs because island biogeography theory relates extinction to the consequence of small population size rather than interspecific competition. Here we need to consider the concept of a minimum viable population size. This is the fewest number of individuals required to sustain a species, and is thought to be around 1,000 for many animals. Once there are too few individuals in a population, in-breeding becomes a problem, which is bad news if a disease occurs because a broad genetic diversity usually ensures that sufficient individuals can survive an epidemic. Small, local populations are also susceptible to small-scale disasters, like a forest fire or flood, which are of little concern to an abundant species spread over a vast area. A further problem arises for rare species because they will have problems finding a mate, thus lowering their ability to replace any losses with new offspring.

There have been few direct studies of extinction but the loss of the heath hen (*Tympanuchus cupido cupido*) from the eastern United States of America (USA) provides an instructive example of what happens when a species' population size becomes critically small. Abundant when Europeans first began colonizing the eastern seaboard, this easily caught and tasty bird was intensely hunted. As a result, by the 1840s it had been wiped out in much of its former range. The trend continued and by the 1870s the heath hen had become restricted to Martha's Vineyard, an island offshore of Massachusetts. By 1896 its numbers had dropped to around a hundred and (rather belatedly) in 1908 a refuge was set up and protective measures put in place. It was too late. Initially numbers rose to around 2,000 but then several harsh winters, a heath fire, and disease that struck what was a genetically in-bred population all combined to wipe out the birds. The last individual, named Booming Ben, was on his own from 1928 and he died in 1932. Clearly if there had been other heath hen populations elsewhere, then the local disasters in Martha's Vineyard would not have spelt the end of the species.

While a useful concept when defining endangered species, the minimum viable population size is of limited use in conservation practice because it is only in hindsight, when a species has gone extinct, that the threshold can be retrospectively judged to have been crossed. Also, conservation efforts show that species can come back from the brink of extinction even when population size is reduced to below 1,000. The North American bald eagle (*Haliaeetus leucocephalus*) provides a nice example. Its numbers are estimated to have fallen from half a million in the early 19th century to around 800 in the 1950s. Today there are more than 100,000 of them thanks to a ban on hunting, the principal cause of their decline. A similar story of success has played out for the trumpeter swan, also in North America. The key point here is that the rarer a species becomes the greater its chance of extinction, but it need not be inevitable even when population numbers become very small.

The relationship between habitat area, population size, and extinction risk allows those species with a high extinction risk to be identified. Thus, endemic species, defined as those restricted to small areas such as islands, lakes, and rivers are always at high risk. Animals near the top of the food chain are usually large and tend to be rare, which increases their risk of extinction. The ability to reach and colonize habitats (known as dispersal), is also a fundamental control on the amount of potential habitat areas a species can occupy. 'Weedy species', like dandelions, produce lots of small seeds that can be blown great distances by the wind. In contrast, many trees rely on large seeds that do not spread far and so have a much more localized distribution. The Californian Joshua tree (*Yucca brevifolia*) is of particular concern because its dispersal speed is thought to be only 1-2 metres/year making it slow to recolonize areas after extirpation (i.e. the local extinction of a species). The Joshua tree is unlikely to cope well with rapid climate change where rapid migration is required for a species to track its favoured climatic belt.

Given the array of factors controlling extinction risk, it might be thought that very widespread and abundant species, such as ocean plankton, would be nearly immune to extinction thanks to their vast populations. However, the fossil record shows us that this is not the case; planktonic species have come and gone despite their enormous oceanic range. This is because the oceans are not as homogenous and endless as they may appear to us. Factors such as water temperature, stratification, mixing rates, and nutrient supply are all mutable ensuring that oceanic habitats can disappear just as terrestrial ones can. Environmental changes and habitat loss have been commonplace over the geological past but currently humanity is the main agent of habitat change (or 'destruction' as it is usually called). This allows us to examine the process of extinction at first hand.

Habitat destruction

Humans have been chopping down trees and making space for farmland and homes for thousands of years. In some regions, such as northern Europe, this process is almost complete and few original forests remain. The main issue and concern today focuses on the destruction of still-extensive tropical forests for their timber and to make room for agriculture and growing populations. These forests harbour a large proportion of global biodiversity and it seems inevitable that, as they disappear, huge numbers of species are going extinct. However, biologists are rarely if ever present to monitor the losses as loggers do their work and so most estimates of extinction losses come from application of the SAR.

One of the few empirical studies of species loss from deforestation comes from Singapore because unusually the species diversity was known before the island's forest were chopped down. Today less than 5 per cent of the original 540 square kilometres of forest remains in a series of small reserves. In total, 28 per cent of the original species have been lost, with birds the worst hit (two-thirds have disappeared), while reptiles and amphibians have been relatively unaffected with only around 5–7 per cent of species lost. Given the scale of the deforestation these losses do not seem too severe, and are well below the number predicted by the SAR. This may be partly because the destruction has occurred recently and there is a considerable 'extinction debt' still to pay. This concept, also known as 'extinction lag', considers that there is a delay between loss of habitat and the death of the species that lived there. Individual animals and plants can persist for several years even though their species is below a viable population size. Survivors are often mature adults who are tougher (with lower mortality rates) than juveniles of the same species, allowing them to live out their life; but the failure of their offspring dooms the species to extinction. For example, the Singapore forests harbour several species of large tree, but with just a few examples of each remaining they are unlikely to produce future generations. As a result, it is estimated that over the next few decades the original diversity of Singapore's forest may be reduced by half due to extinction debt.

Madagascar provides another example of intense deforestation although this time we do not know the original diversity. As little as 10 per cent of the original forest cover now remains as small fragments, and even this has not been preserved in pristine condition. This is especially the case at forest margins. These trees experience a different climate to inner forest areas because the environment there is windier, drier, and exposed to greater temperature fluctuations. As a result of these edge effects, which become proportionally more important as forests decrease in size, the remaining true habitat is even smaller than a simple measurement of forest area suggests. Total extinction losses in Madagascar are unknown, but the well-known lemur family has lost seventeen species in historical times with nearly all the remaining ninety or so species considered to be under threat of extinction.

The SAR can also be used to predict future extinction losses using estimates of current rates of deforestation and habitat loss. Extinction estimates are typically given for either 2050 or 2100, and they vary widely. Thus, it has been suggested that between 5 and 9 per cent of species in the Amazon may be 'committed to extinction' (a phrase that includes actual extinctions plus the extinction debt effect) by 2050. Higher values are obtained when the need for species to shift their distribution due to climate change is also included in calculations. A 2004 study of species losses caused by a combination of habitat destruction and climate change predicted that by 2050, 18–35 per cent of all terrestrial species would be extinct, with the larger number being caused by higher levels of assumed global warming. From a 2004 perspective, over 30 per cent of the time to 2050 has now elapsed and extinction losses have yet to get anywhere near these predicted catastrophic levels. This suggests that using SAR to realistically determine future extinctions may be considerably over-predicting true losses; alternatively the severity of the extinction debt may currently be grossly under-estimated.

Part of the problem, when predicting extinction intensity, is that it is very difficult to account for the proportion of local (endemic) versus widespread species. This is illustrated by the muted extinction levels of birds following widespread deforestation in the eastern USA in the 18th and 19th centuries. Only four extinctions out of a total diversity of ~220 species seem to have occurred. Habitat loss was roughly 50 per cent over this time (although many forests have grown back since) and assuming a z value of 0.25, a commonly used figure, the SAR predicts that extinction losses should have been 16 per cent (thirty-five species). The failure to lose thirty-one extra species may be because the birds have a much broader habitat range than just forests. Even more importantly, nearly all the birds have a much greater range than just the eastern USA. Many of them are also found in the relatively undisturbed forests of Canada, for example; so the intense deforestation only occurred in part of their range. The SAR approach is best used to predict the losses among endemics, but this requires a knowledge of species ranges which is rarely available.

Invasive species

As well as changing habitats, humans have also been responsible for both deliberate and inadvertent introduction of alien species into new areas. Occasionally these immigrants are highly successful and their impact on native species raises frequent concerns. Invaders are a particular issue for island dwellers because of their small population size and limited habitat areas. Also island species often lack an innate wariness making them especially vulnerable to introduced predators. This is particularly the case for many species of Polynesian land snails belonging to the genus *Partula* which have disappeared due to the introduction of a predatory snail (*Euglandinia rosea*). The whole story is one of woeful ecological mismanagement. Originally the giant African land snail was released on the island of Moorea in the hope that it would thrive and provide a food species. It certainly did well but unfortunately it destroyed many crops. A way of getting rid of this new pest was needed, and so a predatory snail was introduced onto the island. Unfortunately, *Euglandinia* found it easier to eat the small, endemic snail species rather than the giant invader which they tended to avoid. Sadly, exactly the same sequence of events has unfolded on several other Pacific islands—release giant African land snail; snail gets out of control; introduce a predatory snail to kill giant snail; predator eats local species instead; endemics go extinct; giant land snail thrives. The death toll of island snail species is probably in the hundreds.

The link between the introduction of a predator and a subsequent extinction is often clear-cut, as in the case of island snails. A survey of recent extinctions among birds, mammals, and reptiles found nearly 150 examples where predation (especially by cats) was the clear cause. Flightless island birds are especially vulnerable, for obvious reasons. However, it is not just predators that have been introduced. Many plants have been moved around the world, either as farm crops or just for their ornamental appeal. *Rhododendron porticum* was introduced into United Kingdom gardens in the 18th century and has gone on to become a major pest, transforming diverse habitats into *Rhododendron* thickets. It has had similar success in New Zealand. What is not clear is whether such plant introductions, which cause problems by filling habitat space rather than by directly killing endemic species, can indeed cause extinction on their own. The remnants of Madagascan forests are under particular pressure from introduced plant species, for example, suggesting it is of serious, albeit secondary concern compared with the primary issue of ongoing deforestation.

It is not just plants and animals that have been moved around. Amphibians are currently suffering all over the world from a fungal infection, called chytridiomycosis, which is thought to have been spread by the transportation of frogs in aquariums. The specific vector may have been the African clawed frog *Xenopus* which is widely used in medical research and thus traded throughout the world. Many frog species have already gone extinct as a result of the disease and vastly more are threatened.

In some cases, invasive species may be falsely blamed for extinctions. A case in point is the extinction of freshwater unionid bivalves in North American lakes and rivers, which has been attributed to the introduction of European zebra mussels (*Dreissena polymorpha*). These mussels attach themselves to the unionids, prevent them from burrowing and thereby hindering their filter feeding. However, the zebra mussels are a recent invader that only became widespread in the 1990s and the unionid extinction process has been ongoing since the start of the 20th century. Environmental degradation, such as pollution from fertilizers, or the total loss of habitats due to dam building, are probably the key factors while the role of zebra mussels is likely a late-arriving secondary stress. A similar debate relates to the introduction of the Nile perch (*Lates niloticus*), a large predatory fish, into East African lakes in the 1950s. This was followed by a rapid demise of numerous endemic cichlid fish species (with a much clearer cause-and-effect timeline), but other factors, such as pollution, may also be important.

Hunting and fishing

Even if the Nile perch has enjoyed success by gobbling up endemic cichlids in African lakes, it is now facing its own problems: its abundance has fallen dramatically in the 21st century. This is due to intense fishing which, along with hunting, provides one of the most common causes of extinction in recent times. Over-exploitation, as it is more generally known, is a species extinction process that has no parallels in nature or in the geological past. Often species are hunted for food but sometimes the extermination is just for the fun of it. Thus, the famous dodo (*Raphus cucullatus*) was reportedly unpleasant to eat but it was very easy to kill. For bored 17th-century sailors given shore leave after months afloat, the chance to bludgeon dodos to death with sticks at least provided a sort of entertainment, it seems. The introduction of cats and rats to the island probably did not help. Island-dwelling birds have been especially vulnerable to extinction. It is estimated that the gradual spread of humans across the Pacific islands over the past 3,500 years has caused the loss of a third (~450) of their species.

Island species are clearly the most vulnerable to extinction due to their low population numbers and limited geographic range. However, they are not the only ones to have suffered at the hands of humans; even some quite widespread species have been wiped out. The Caribbean monk seal (*Neomonachus tropicalis*) was found throughout the Caribbean (as its name suggests), but it was so intensively hunted for food and oil in the 18th and 19th centuries that it finally disappeared in the 1950s. Probably the most spectacular example of hunting-driven extinction is the elimination of the passenger pigeon (*Ectopistes migratorius*) in North America. It is estimated that their population may have exceeded three billion in the early 19th century making it the most abundant bird species in the world. Within a hundred years, however, they were gone: in 1914, the last individual, Martha, died in her cage in Cincinnati Zoo. That hundred years had seen an orgy of hunting the pigeons for food and sport. Legislation was proposed to protect the pigeons, beginning in the 1850s, but it was thought unnecessary at the time because the notion that such a common species could go extinct was an alien one. By the time protective bills were passed in the 1890s it was already far too late. In more recent years, the role of habitat destruction and natural boom-and-bust population cycles have also been implicated in the demise of the passenger pigeon but these are likely to have been of minor importance compared to their hunting, whereby birds were being killed at a rate of hundreds of thousands a week, for months at a time.

A much more ambivalently felt extinction, contemporaneous with that of the passenger pigeon, is that of the Rocky Mountain locust (*Melanoplus spretus*). Like many locust species, some years it would occur in such huge swarms that it devastated crops over vast areas. The last such mass occurrence was in 1877, yet by 1902 it was extinct. The intervening years did see intense efforts to eradicate the pest—people were paid to kill them—and, furthermore, it is thought that many locusts were destroyed by intensification of agriculture and ploughing of land where their eggs were found. However, the extinction remains puzzling and a lack of knowledge of the locust's ecology makes it unclear how they disappeared so abruptly. Nonetheless, North American farmers have been happy about their disappearance ever since.

Climate change

In recent years it has been claimed that human-effected climate change will become the number one cause of extinctions this century due to the release of carbon dioxide (CO_2) from the burning of fossil fuels. This greenhouse gas drives global warming and it is predicted that, by the end of the current century, global temperatures may have risen by somewhere between 1°C and 4°C. This will require many species, particularly those with a narrow climatic tolerance, to migrate poleward, a task that will be rendered difficult for many because of habitat fragmentation. In some cases, species will have to 'island hop' across agricultural and urban 'oceans' to find undisturbed 'islands' of natural habitat. The ability to do this will depend on dispersal ability which is poorly known, but for many species it may be quite effective. Thus, recent analysis of several thousand species has shown that this migration is already underway with average migration rates of 17 kilometres/ decade. This has been sufficient to keep pace with warming and there have been few if any reports of extinction caused by climate change so far. This finding is perhaps not that surprising because there have been many large and rapid climate changes over the past 2.5 million years as the Earth has cycled between icehouse to greenhouse phases. The inhabitants of our planet are used to rapid climate change.

Dispersal and migration are much less of a problem in the oceans because habitat destruction is more limited and generally focused around shorelines. However, climate change may still play a role in future marine extinctions. Carbon dioxide dissolves in the surface of the ocean causing a slight decline in the pH—a process called ocean acidification. This has been postulated to be a future threat to marine species, especially those that secrete shells of calcium carbonate, which becomes prone to dissolution as pH declines. The predicted intensity of acidification is only modest though (<0.4 pH units by the end of this century), and no species are yet under extinction threat.

It is in cold waters that ocean acidification will be most intense because the amount of CO_2 that can be dissolved in water increases as temperatures cool and so polar waters will become the most acidified. The pteropods, a group of planktonic snails abundant in polar waters, are considered particularly at risk with the Intergovernmental Panel on Climate Change (IPCC) predicting that they will be extinct by 2100. They secrete a tiny skeleton of aragonite (a relatively soluble form of calcium carbonate), a millimetre or so in size, and flap around in surface waters eating plankton. In their turn, the pteropods are eaten by many species, including fish, and they form a key part of polar food chains. Their extinction would have dire consequences for high-latitude fisheries but, twenty years after the first warnings, the pteropods are not yet in peril.

Rather than acidification, temperature increase is a more likely cause of stress for many marine species as shown by the impact of episodes of extreme warmth on tropical reefs. These are constructed by corals which harbour green algae in their tissues (from which they derive nutrients). The algae, called zooxanthellae, die if temperatures approach 30°C causing the corals to appear 'bleached'. If temperatures return to normal, then the corals can regain their algae but long-term bleaching is damaging to the health of the corals and reefs in general. Reefs harbour diverse communities and it is possible that prolonged bleaching episodes will drive some species to extinction as reefs crumble.

The fate of modern reefs, like that of many terrestrial habitats, is also tied up with direct action by humans. Many reefs are suffering from over-fishing and environmental change caused by pollution and increased run-off of sediment from land caused by deforestation. The direct actions of humans is less obvious in polar latitudes where increased temperatures are likely to be especially harmful with these regions being predicted to experience the greatest change. Also, while many habitats will migrate polewards, as temperatures rise, the highest latitude habitats will either shrink or disappear entirely.

The IPCC has the job of predicting the future climate for policy makers. Ostensibly they have an easy task because the long-term consequence of greenhouse gas emissions is clearly going to be global warming. This conclusion is supported by experimental data and our knowledge of the many CO_2-driven warming episodes in the geological past. In the short term though climate change is very hard to understand because of the vagaries and variations of climate and weather on a decadal scale. This puts the IPCC in a bit of bind: if they were to claim that the climate will be a lot warmer in 1,000 years' time as a result of fossil fuel burning, then few would doubt them. Unfortunately, politicians would not pay much attention to such statements. To get their message heard the IPCC is therefore forced to give shorter term predictions about the consequences of climate change within this century—a timespan of more urgent importance to the Earth's current inhabitants. The down-side of this approach is that we only have to wait a little while to see if they come true, and if they do not then the Panel loses some of its credibility.

The IPCC was founded in 1989 and it has produced regular reports ever since, on both future climate change and extinctions. With hindsight most of their earlier claims have yet to be borne out. A 2007 report claimed that up to 70 per cent of the world's species will have been driven to extinction by global warming by the end of this century with half of the losses achieved by 2050. Over a dozen years later there is no sign of extinction losses of this magnitude, and the 2014 IPCC report reduced their species extinction estimate to <30 per cent, suggesting that 80 per cent of these losses would be due to habitat loss (and therefore unrelated to climate change). This is a major change of emphasis in just seven years. More recent reports have not even reiterated the 2014 prediction and instead have focused on temperature-sensitive communities. In October 2018, IPCC representatives stated that global temperatures are likely to increase by 2°C before 2050 and cause the extinction of more than 99 per cent of the world's coral reefs, with losses anticipated to begin within the next decade. Other extinctions were said to be likely, due to factors such as deforestation and hunting. The latter claim is no doubt true, but these are not climate change issues.

Extinction today is being demonstrably caused by human activities such as habitat destruction, the introduction of invasive species (especially predators), and the general over-exploitation of natural resources. These losses have been especially intense among vulnerable island species with their small habitat range and population numbers, although freshwater systems (lakes and rivers) are also at risk. Future climate change is clearly an additional potential cause of extinction. Many of the likely causes such as warming and ocean acidification are often held responsible for past crises which lends credence to our concerns for the near future. Knowing when these changes will begin to take effect is difficult to predict, but the IPCC believes we only have to wait a few decades longer.

Chapter 2
Extinction today and efforts to stop it

A sixth mass extinction

The fossil record shows that life has experienced five major catastrophes, known as the 'big five' mass extinctions, and a sixth catastrophe may be underway. This latest crisis is variously called the sixth mass extinction, the Holocene extinction (named after the interval of geological time in which we are living), and the Anthropocene mass extinction (named after a newly proposed interval, spanning the last few centuries, when human's environmental impact has been at its most intense). The mass extinctions of the past were geologically short-lived, intense crises that affected animals and plants in all environments on both land and sea. They removed the dominant and abundant species in an environment, often called the incumbents (e.g. the dinosaurs), and left an ecological void to be filled by groups that were often rare or insignificant beforehand (such as the mammals before the dinosaurs). Uniquely, the big five also saw the collapse of the base of the food chain in the oceans and the extinction of many planktonic groups. The current extinction crisis does not yet have any of these attributes. Losses have not even begun to approach those which occurred during the mass extinctions, neither are the current extinction losses seen across the complete spectrum of environments (there is no extinction crisis in the deep ocean, for example) and photosynthesizing plankton remain in robust health. Instead, the observed species losses are mostly concentrated among large-bodied terrestrial vertebrates with small populations.

The International Union for the Conservation of Nature (IUCN)—a conservation body dedicated to monitoring and preserving diversity—maintains a 'Red List' of endangered species and also those that have recently gone extinct. This shows that there have been around 1,200 recorded extinctions in the past few hundred years, with most losses found among island faunas. Common species have fared much better, with occasional exceptions like the passenger pigeon. Furthermore, while the big five mass extinctions wiped out huge numbers of marine invertebrates only twenty marine species appear to have gone extinct recently.

In many regards, the claim for a sixth mass extinction fails to meet the criteria that define this term because extinction losses so far have been minor and highly selective. Nonetheless, current extinction rates may be comparable, suggesting that we may be witnessing the start of a mass extinction episode. Normal levels of extinction, known as the background rate, are usually stated to be between 0.1 and one species extinction per 10,000 species every hundred years, which can also be stated as an extinction rate of 0.1–1.0 species extinction per million species per year, a metric known as E/MSY. These estimates are calculated from fossil data on extinctions among marine invertebrates which are generally widespread and abundant. These factors are likely to have reduced the extinction risk, and so the rates derived from marine invertebrate data may be on the low side. Many terrestrial animal groups in particular seem to have higher extinction rates than the invertebrates. Thus, background rates for mammals over the past few million years is 1.8 E/MSY. It is important to bear this number in mind because the current extinction rate for mammals is among the best known and so it provides the best comparison with past rates.

Since the year 1500, mammal extinction rates have averaged 14 E/MSY, based on IUCN data, which is clearly significantly elevated above background. Furthermore, most of the losses have occurred in the past two centuries, and since 1900 the rates have risen to 28 E/MYR. Similar rate increases are seen among other vertebrate groups (Figure 1) and these data provide us with clear evidence that there is an ongoing and severe extinction crisis. Whether a 'mass extinction' epithet is appropriate is a moot point though, because these losses are just among terrestrial vertebrates. Our lack of information on the fate of other groups makes it difficult to know if the extinction crisis is more widespread. For example, terrestrial biodiversity is dominated by the insects, and yet we have little idea of their extinction losses, except for within the best known groups, like moths and butterflies; and they too have high extinction rates like the mammals.

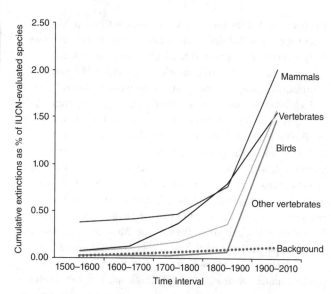

1. Graph showing the acceleration of extinctions over the past 500 years among terrestrial vertebrates based on recorded extinctions by the IUCN.

Extinction rates may be even higher than estimates suggest because they are based on observations of animals that are either large, and therefore easy to survey, or otherwise clearly visible, like birds and butterflies. We know relatively little about the fates of small, rare, and obscure species. It is estimated that only a third of the total number of species on the planet have been described, with many of the unknown ones being hard to find because they are rare and/or have small geographic ranges. These are precisely the attributes that make a species prone to extinction and so it is probable that actual extinction rates, when including unobserved species, are much higher than we realize. An alternative to reliance on the incomplete IUCN lists is to calculate extinction rates using the SAR and known habitat loss. This approach suggests we are losing between 11,000 to 58,000 species annually, and it produces much higher estimates of current extinction rates: somewhere between 1,000 and 10,000 times the background rate.

So are the current extinction rates—however poorly they are constrained—comparable with those of the past? This question is not very relevant to our understanding of the modern crisis because of its unique nature, and although it is often asked, the truthful answer is that we have little idea. Here's why. Extinction rate calculations vary enormously depending on how long an extinction interval is thought to have lasted, and so there are no reliable E/MSY for past mass extinctions. For example, up until the early 1990s the Permian-Triassic mass extinction—the greatest crisis of the fossil record—was generally thought to have been spread over many millions of years. However, more recent research indicates it occurred over a much shorter duration. Even so, current estimates range from 30,000 to 200,000 years. With such a broad range of extinction times to choose from, calculated extinction rates vary enormously. In another example, a recent study compared modern extinction rates with those during the mass extinction sixty-six million years ago that eliminated the dinosaurs. This suggested that current diversity losses may be up to 18,000 faster. This is an impressive figure but it was achieved by assuming that the dinosaurs' demise was spread over somewhere between 0.5 and 2.75 million years. This is a strange choice of time range. It is much more likely that the losses happened within just a few months and years of a giant meteorite impact in Mexico, making it an interval characterized by the highest extinction rates ever known, certainly vastly greater than those today.

There is also an apples-and-oranges-style problem when comparing ancient and modern extinctions. Past mass extinction data are dominated by marine losses (supplemented with less well-known terrestrial losses), whereas modern data are almost entirely based on terrestrial extinctions because most of the losses seem to be happening on land. So it is hard to quantify and compare rates and styles of extinction between those in the past and those in the present. With historical extinctions we have a relatively good handle on the time frame but many species losses are unrecorded, whereas while past extinctions among, for example, shelly marine invertebrates are well-known, the timespan of the extinctions is not.

Terrestrial conservation

Whatever epithet is applied to the current extinction crisis, there is no doubt that global biodiversity is under severe threat from human action. A variety of distinct conservation approaches seek to stem these losses, and many schemes have been underway for decades; some have met with modest successes. Eradicating invasive species, especially cats and rats on small islands, has been particularly beneficial for endemic species and is a relatively small-scale and cheap conservation approach. On the whole, conservationists are facing a rising tide. In 2002, the parties to the Convention on Biological Diversity (CBD)—a United Nations-sponsored organization representing nearly all the world's countries committed to significantly reducing the rate of biodiversity loss by 2020—proposed a series of measures to achieve this. A 2014 analysis showed that essentially none of

CBD's targets were even close to being met, although there was an increased global awareness of the need for conservation.

On a large scale the twin drivers of extinction, habitat loss and hunting, are combated by designating protected areas (e.g. national parks, game reserves) and by limiting or banning hunting. Thus, since the 1960s the Convention on International Trade in Endangered Species of Wild Fauna and Flora (CITES) has attempted to protect some of the world's most endangered species by restricting their sale around the world to help sustain populations. It has had some considerable success but the economic value of species is often their downfall. The problem is illustrated by the ongoing fate of elephants. They are killed for a number of reasons—to protect farmland or for sport—but principally it is their tusks that are the main target because ivory can be carved into ornamental objects. Nowadays, the principal ivory market is in China and Japan where it is used to make *chops* or *hankos*, respectively—carved cylinders with their owner's seal used for stamping documents. Sadly, as the wealth of the population in Southeast Asia has increased over the past decades so has the demand for these ivory stamps, putting inevitable pressure on cash-strapped African nations to meet demand. CITES attempted to help in 2008 by overturning the ban on elephant hunting, which had been in place since 1990, to allow some African nations to sell their stockpiles of ivory and to cull a limited number of elephants Sadly, this action has only served to boost the trade in illegal (and legal) ivory. CITES has a programme called MIKE (Monitoring the Illegal Killing of Elephants) which has shown that the rate of poaching rapidly increased after 2008 and now greatly exceeds sustainable levels. In 2016, the total population of the African elephant was estimated to be 415,000, down from 526,000 in 2000 (and a small fraction of the population in the 1940s, estimated at five million). With poaching rates of around 30,000 per year, by 2025 there may only be 200,000 African elephants left. There is a glimmer of hope. The Chinese government is beginning to recognize the damage the ivory trade causes, and in 2015 it imposed restrictions (but not a ban) on ivory imports.

The rapidly increasing wealth of Southeast Asian populations over the past three decades has generally had a detrimental effect on some of the world's rarest and most endangered species. For the most part, these species are targeted for traditional Chinese medicines, although the medicinal value of the products is generally based on superstitious beliefs rather than scientific evidence. Thus, for example, powdered rhino horn is claimed to treat a broad range of ailments including rheumatism and headaches even though it is simply made of keratin—the same material found in horses' hooves and our finger nails; the scales of the pangolin or scaly anteater are also made of keratin and these too are thought to have medicinal use—as a result, Asian pangolins are now critically endangered and African pangolins are being hunted at an estimated rate of 2.7 million per year. Belatedly, CITES imposed a total ban on the pangolin trade in September 2016. For some species, their supposed medicinal properties have already led to their extinction. The kouprey, a large forest-dwelling buffalo from Southeast Asia, was hunted because its horns and skull were ground to powder for use in Chinese traditional medicines. The unfortunate animal has not been seen for over thirty years and is almost certainly extinct. Beliefs in the medicinal properties of animal parts are not restricted to Southeast Asia. In Tanzania, it is thought that eating the bone marrow and brain of giraffes provides a cure for AIDS. As a result of their hunting (and also habitat loss), giraffe populations are in rapid decline.

When a species becomes very rare it often becomes necessary to capture the remaining individuals and instigate a captive breeding population. For example, the Sumatran rhino was declared extinct in the wild in 2015 with the hundred or so individuals that remain cared for in reserves and zoos. If population sizes increase sufficiently, then it is sometimes possible to reintroduce species into the wild, although such strategies have a poor success rate

(<25 per cent) because the low genetic variability of small, in-bred populations makes them vulnerable to disease. Grander schemes of reintroduction, also called 'assisted colonization' in the terminology of the IUCN, or 'rewilding', have also been proposed. The reintroduction of wolves to Yellowstone National Park in the 1990s is one example considered to have been a success. There have even been proposals to use clonal techniques to resurrect long-extinct animals such as the mammoth. However, it might be better to leave these ideas to science fiction writers and spend the money on saving species not yet extinct.

Many long-standing conservation projects prioritize high status, charismatic animals such as the rhino, the tiger, or the poster child of the World Wildlife Fund, the panda, because of their value in fundraising. This approach has the additional benefit of protecting the habitat where the signature species live, and so incidentally helps protect entire ecosystems and many other species. However, there has been considerable debate about conservation strategies and what should be prioritized. One approach would aim to protect the most endangered groups. Top of this list would be the amphibians with over 40 per cent of species under threat of extinction, compared to an average of 20 per cent for vertebrate species in general. The IUCN Red List contains many frogs and toads with magnificent names such as the leopard rocket frog, the skunk frog, and the Mesopotamian beaked toad. Sadly, the list also includes many recently extinct species with equally impressive names: the Mount Glorious day frog (extinct in 1979), the gastric brooding frog (extinct in 1983), the golden toad (extinct in 1989), and Rabb's fringe-tinged tree frog (the last known individual died on 26 September 2016). Globally, amphibian populations are suffering from a fungal infection, but the main cause of their decline is habitat destruction, and their conservation requires the establishment of numerous protected reserves.

An alternative conservation approach involves focusing efforts on protecting unusual species and those with no close living relatives.

This idea is pursued by the 'EDGE of Existence' programme launched by the Zoological Society of London in 2007. EDGE is an acronym for 'Evolutionarily Distinct and Globally Endangered'. The solenodon, an insectivorous mammal found in Cuba, is a typical example of an EDGE species. It is very similar to the primitive mammals that were around at the time of dinosaurs and has no close relatives. It thus represents a very long and lonely branch of the mammalian evolutionary tree. Rather than target individual species, some conservationists suggest focusing efforts on biodiversity hotspots—the places with the most species to lose. These include Central American rainforests, the northern Andes (especially for plants), South Africa (also for plants), the Amazon rainforest (especially for amphibians), and the forests of New Guinea and the Philippines.

To an extent these conservation strategies have overlapping aims and targets (in a nutshell it could be called the 'frogs and forests' approach), but there is also a conflicting strategy that has grown out of the field of climate modelling. Global warming will require species to migrate to higher latitudes to keep pace with their optimum conditions. Similar habitat tracking has happened many times during the glacial and interglacial cycles of the past two million years; but the growth of agricultural land and urbanization will make this process more difficult. So it has been suggested that migration corridors be established. The focus has been in North America because it already has a relatively high density of protected land, with a proposal for the creation of a Yellowstone-to-Yukon corridor. Such an ambitious scheme would be difficult to achieve for both legal reasons (a lot of privately owned land would need to be requisitioned) and the enormous expense. Many conservationists have questioned the value of such approaches.

Much of the world's biodiversity is found in tropical areas which, climate modelling shows, are expected to experience only modest global warming (1–2°C by 2100), whereas the greatest temperature increases are likely to occur in high northern latitudes (6–7°C by 2100) where biodiversity is very low. It is therefore a moot point whether connectedness is a serious issue when the main cause of extinctions is habitat destruction and hunting in the tropics. Estimates of future species losses in the 21st century generally suggest that fewer than 10 per cent of extinctions will be due to climate change. Even the IPCC's 2014 assessment acknowledges that 80 per cent of future losses will be due to land-use change, with the remainder due to a combination of this and climate change.

Marine conservation

Compared with terrestrial extinctions, the losses in the marine world have been much less severe. This observation can be viewed optimistically—maybe marine species are much less prone to extinction; or pessimistically—the extinctions have yet to come. There is no doubt that marine species have an advantage compared to land-tied ones because, with a few exceptions, such as mangrove swamps and reefs, true destruction of marine habitats is near impossible. There is no ocean equivalent of deforestation and urbanization. Also many marine species have a much broader geographic range than terrestrial species—a good insurance against extinction—and their migration is unhindered by farmland and cities. This is true both for fish, which actively swim, and sedentary, bottom-living species (the benthos), which frequently have planktonic larvae that disperse over vast distances.

Despite factors that favour survival of marine species, there is a clear need for marine conservation because over-fishing has major economic consequences and for many people it is an essential part of their livelihood. It is estimated that at least a hundred million people rely on subsistence fishing for a living. Also, over-exploitation has placed many species on the endangered list, particularly those that are slow growing (e.g. turtles, whales) or those with limited geographic range (e.g. some dolphins).

Some species are being inadvertently endangered because they become ensnared in fishing nets and lines (e.g. dolphins, albatrosses) even though they are not directly targeted; for some of these bycatch species, extinction is probably just around the corner. The vaquita porpoise (*Phocoena sinus*) is a typical example. It is restricted to the Gulf of California where it is accidentally killed in the gill nets used to catch the totoaba fish. As an aside, the totoabas are also endangered but they are nonetheless still being intensively fished for their swim bladder, used in Chinese cuisine because it is thought to have medicinal properties. Restrictions on totoaba fishing have been put in place but they are weakly enforced and the vaquita population has consequently plummeted from 300 in 2011 to fewer than thirty in 2017. Plans are afoot to use specially trained dolphins to find the last remaining vaquita so that they can be captured and their numbers increased in captivity. A similar fate is also playing out for the Maui dolphin (*Cephalorhynchus hectori maui*) found in the waters around North Island, New Zealand. It is also caught as a bycatch by fishermen, and government restrictions—only put in place in 2003 when a hundred were left—have proven ineffective. The current population has fallen below fifty individuals.

Marine conservation efforts usually come late in the day for most species, often when their numbers have been reduced to a small percentage of their original population size. At this point the economic cost of finding and catching rare species becomes excessive and so most countries are willing to put in place fishing bans because there is little economic counter-argument. Some species subsequently recover whereas others remain moribund, although surprisingly few over-fished species have yet gone extinct. The history of whale hunting provides a case example.

Whales have been hunted for hundreds of years, and one species—the Atlantic grey whale—was driven to extinction before the 18th century. However, the intensity of whale hunting increased enormously in the 20th century as demand for their oil grew, and ships and hunting techniques became more effective. By the early 1930s, 30,000 whales a year were being killed and several species had become very rare. Usually the largest species were hunted first and then, as they became harder to find, smaller species were exploited. This trend can be seen in the mismatch between whales killed and the oil yielded. Thus, in 1933, 30,000 whales produced 2.5 million barrels of oil while, in 1967, 60,000 whales were killed but only half the tonnage of oil was obtained because, by this time, small whales like the minkie (*Balaenoptera*) species were the main target.

Efforts at conserving whale stocks began in the 1930s when the League of Nations imposed a ban on hunting some of the rarest species, although making little effort in enforcement. The International Whaling Commission (IWC) was set up in 1946 with the aim of making hunting more sustainable, but it was met with little success or enthusiasm. A 1972 call by the United Nations for a ten-year ban on whale hunting was ignored by the IWC. However, by this time, numbers were so low that the whaling industry had essentially self-destructed: only Japan and the USSR still had ocean-going whaling fleets by 1980. Furthermore, whale oil substitutes made from petroleum were being widely used, weakening the commercial value of whales. By 1982, there was little economic reason to oppose the complete ban on whaling proposed by the IWC. This was enacted in 1986 and has been in place ever since, although Japan (and a few other countries) still hunt a few whales, ostensibly for scientific investigation. Over thirty years later the benefits of the ban are seen in the growing populations. Humpback whales (*Megaptera novaeangliae*), reduced to a few hundred in the 1950s, now number in the tens of thousands. Other species have been much slower to recover probably because of extreme rarity and slow reproduction rates. The blue whale (*Balaenoptera musculus*) population had been reduced to a few hundred individuals, but it has increased at a rate of a few per cent a year to around 15,000 in 2016.

The story of whale hunting is being re-enacted today in the tuna industry. Tuna are large, fast-swimming fish found throughout the world's oceans where they are important top predators. Globally tuna fishing is thought to be worth £29 billion and most fishing is considered to be unsustainable (i.e. populations are in decline). As with whales, the larger species were initially the main targets, with the result that their populations are the most depleted. Thus, the Pacific bluefin (*Thunnus orientalis*) population has shrunk to 3 per cent of the pre-fishing population, and is listed as threatened with extinction by the IUCN. In contrast, the much smaller skipjack tuna (*Katsuwonus pelamis*) is globally distributed and not considered under threat at the moment, even though nearly three million tonnes are landed each year. Tuna currently lack an equivalent to the IWC, instead there are organizations with a regional remit, like the International Commission for the Conservation of Atlantic Tuna, which sets (unenforced) limits on catches. The Inter-American Tropical Tuna Commission has made some minor, but more successful, attempts to close some fishing areas and impose quotas on the nations, such as Japan, which harvest the lion's share of the ocean's tuna. Tuna fishing has currently reached the situation seen with the mid-20th-century whaling industry: large species have become very rare, while small species are not yet threatened.

The fishing industry in general grew rapidly in the latter half of the 20th century, and by 1996 total landings were 130 million tonnes. Since the start of the 21st century this number has been falling (it was 109 million tonnes in 2016), a clear indication that fishing intensity is unsustainable. There has also been a change which has seen fishing expand to progressively deeper levels (commercial fishing today extends down to a depth of nearly 1.5 kilometres) because shallower stocks have become severely depleted in many regions. A spectacular example of what over-fishing can do was seen in the Grand Banks Fishery on Canada's east coast. One of the main cod (*Gadus*) fisheries in the world, whose catches quadrupled in the 1960s to reach over

700,000 tonnes per year in the early 1970s, rapidly collapsed to zero in 1992, forcing 35,000 people into unemployment in the Newfoundland fishing industry. In the same year, but much too late, the Canadian government imposed a moratorium on fishing in the region. Despite the ban, there has been little recovery in cod numbers. It appears a fundamental change has occurred in the ecosystem, one that no longer allows cod to thrive.

Governmental efforts to maintain fishing stocks have met with variable success. One of the most effective has been the Sustainable Fisheries Act passed by the US government in 1996. This introduced measures such as closure of fishing grounds and scientific monitoring to maintain fishing at sustainable levels. The result has seen commercial landings remain stable along with the number of people employed in the US fishing industry. In contrast, the European Union's (EU) fisheries policy has failed to follow scientific advice on sustainable fishing levels, with the result that landings have declined 21 per cent between 2000 and 2015 (with a concomitant decline in the size of the EU fishing fleet). In an attempt to alleviate this problem, the EU negotiated fishing rights for European vessels in the territorial waters of Mauretania. By paying compensation of €59 million/year, EU vessels are allowed to catch 280,000 tonnes/year, an amount considered, however, to be unsustainable by activist groups like Greenpeace. The EU calls the practice a 'Sustainable Fisheries Partnership', while Mauretanian fishermen probably use a different phrase.

As a final word on global fishing—while it has yet to achieve much in the way of extinctions it has nonetheless rapidly and fundamentally changed the nature of marine ecosystems. Large predatory species, like tuna and sharks, are the most fished and depleted stocks. As a result, marine food webs have been fundamentally altered in the past hundred years in a way that has no parallels in the geological past. We have little idea what the short- or long-term consequences of decapitating the top of the food chain are, both for marine life and our future ability to harvest the oceans.

In contrast to the rest of the marine realm the conservation problems of reefs are much more akin to those of terrestrial habitats because they represent an environment that can be destroyed. The main threats come from over-fishing, acidification, coral bleaching during warm years, and siltation due to sediment run-off from the surrounding land often caused by deforestation. Pollution, caused by influx of nutrients from adjacent farmland is also a serious problem for the health of reefs because they are generally found in waters with very low nutrient supply. Many species of coral symbiotically farm photosynthetic algae that supplements their diet and allows them to grow in low-nutrient conditions. However, they get out-competed by the growth of seaweed if nutrient levels increase. Added to these issues, many reef species have limited geographic range, making them vulnerable to extinction if their habitat is changed. Some have suggested that a quarter of the world's coral species may disappear by 2050 due to these various causes, while a recent IPCC report claimed global warming will wipe out all the world's reefs by 2050.

Protecting and conserving the world's reefs is clearly a major challenge, and if they are vulnerable to global warming then there may be little that can be done. More direct causes of damage, such as intense fishing, can be alleviated by designating no-fishing areas. This allows spawning to occur unhindered and restocking of adjacent areas with fish—a sustainable solution to the benefit of subsistence fishing industries that characterize many of the areas where reefs occur.

Chapter 3
Extinction in the past

Reading the record

The majority of fossil species are no longer living, thereby providing ample evidence that extinction is a frequent phenomenon. However, before we assess the patterns and processes of extinction it is important that the imperfections of the fossil record are understood. The most obvious bias is the preferential fossilization of hard parts (bones and shells), while soft parts are rarely found. Thus, shelled organisms, like molluscs, have an exceptionally good fossil record whereas entirely soft-bodied organisms such as worms do not, ensuring that the former have an apparently more continuous time range. Equally important, sedimentary rocks do not provide a continuous record of geological time because they contain surfaces that represent periods when sediment was not deposited. Thus, the history of life is full of hiatuses, and the range of most species typically consists of a series of intermittent occurrences. Darwin, in his *Origin of Species*, was the first to recognize the importance of this phenomenon and how it may cause extinctions to apparently cluster at certain horizons in the fossil record (Figure 2). These represent levels with a lot of time missing during which many species disappeared. Thus, an apparent mass extinction can actually be caused by prolonged time gaps in the fossil record. This problem is generally at its

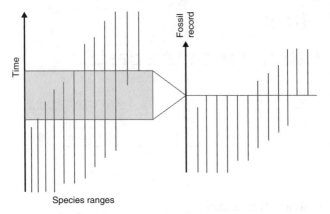

2. Conceptual figure showing how an apparently sharp extinction event in the fossil record can be created by the presence of a major time gap (hiatus—shown shaded) in the geological record.

worst for terrestrial and shallow marine rock successions because they tend to be very 'gappy'. It is much less of an issue for deep marine sedimentary rocks because they accumulate more continuously.

While time gaps can create apparently abrupt extinctions, the poor fossil record of rarer species can conversely create an apparently gradual extinction when in truth there is an abrupt, mass extinction (Figure 3). This phenomenon has been named the Signor–Lipps effect, after two palaeontologists who first highlighted the issue, or sometimes it is called the less pleasant-sounding 'back-smearing effect'. It is caused because the last/youngest fossil found is very unlikely to have been the last individual to have lived. Thus, apparent/observed extinction occurs earlier than the true extinction. This is especially so for uncommon fossils which tend to have a very patchy occurrence. We can see this in our understanding of dinosaur extinction and its relationship to a meteorite impact in Mexico at the end of the Cretaceous.

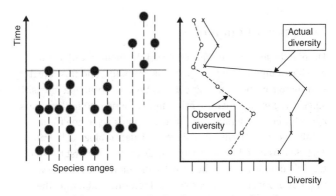

3. Conceptual figure showing the fossil occurrences of species as a series of points. The dashed vertical lines show the true time range of species and it will be seen that the disappearance of rare species (i.e. those with few occurrences) is often long before their actual extinction.

Dinosaurs are rarely found as fossils, and the youngest examples come from below the level of impact. There are no dinosaur bones known to be sprinkled in meteorite dust. Thus, during the 1980s, many palaeontologists claimed that the demise of the dinosaurs had nothing to do with a meteorite because they had already faded away and were on the point of extinction before the impact. However, there has been much intense fossil-hunting activity in the last thirty years which has gradually led to the last occurrences of many dinosaur species being raised nearer to the impact horizon. With cause and effect much closer together the extinction–impact link is now a lot clearer. In fact, even in the 1980s, this connection was well-established for very common species because their disappearance provides a much more accurate indication of extinction (i.e. the Signor–Lipps effect is less important). Foraminifers are tiny shelled protists which are part of the plankton. Their name is usually shorted to 'foram' and they have offered the best fossil record in the Late Cretaceous because their shells occur in prolific abundance in deep marine sedimentary rocks. At the end of the Cretaceous nearly all foram species abruptly disappear at the impact level, a sure sign that the extinction was abrupt.

Extinction through time

Despite the less than perfect nature of the fossil record it still provides a unique window on the history of life, and this reveals that there have been dramatic fluctuations in extinction intensities since complex life (metazoans) evolved around 600 million years ago. Our best understanding of these trends has come from database compilations taken from the palaeontological literature. Jack Sepkoski of the University of Chicago was the first to attempt this in a detailed way in the early 1980s and he spent many years in libraries reading palaeontological literature and noting the occurrences of fossils through time. He compiled this information to produce a series of highly informative charts showing both diversity (Figure 4) and extinction rates (Figure 5) since the start of the Cambrian Period 541 million years ago.

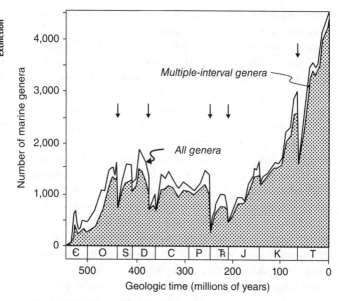

4. Global marine diversity during the past 600 million years showing the steps caused by the big five mass extinctions (arrowed).

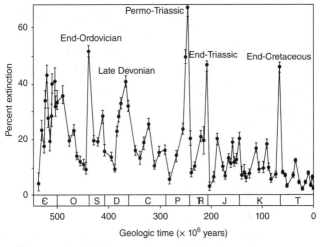

5. **Phanerozoic extinction rates for marine genera.**

Sepkoski's initial range data were for marine families, and later he enlarged his database with a compilation of marine genera. Sepkoski had to compromise at this taxonomic level because it would have been a vast undertaking to compile a species database (in any case it took him many years of work just to get the family and generic data). In fact, there is little point in compiling species extinctions because the quality (patchiness) of species occurrence is much worse than for genera. Many species are often only found in a single location, giving the impression that they appear and disappear (i.e. become extinct) at a moment in time. Such single-point occurrences, as they are known, are caused by the imperfect nature of the fossil record. Genera consist of several species and so they have a greater chance of turning up as fossils, ensuring that their range through time is better known.

Despite the lack of information, it is still possible to evaluate species extinction percentages. These values have usually been calculated using generic extinction data and graphs that show the

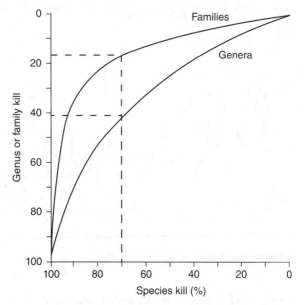

6. Graph showing the relationship between genera and family extinctions to species extinction rates.

number of species versus the number of genera (or families) going extinct (Figure 6). These plots are based on the hierarchy of the classification system we use for organisms. Species are assigned to genera; these in turn belong to families; they to orders; and so on. As a result, there are many species, fewer genera, and even fewer families. This means that if one species goes extinct then there is a good chance that its genus will still survive because there will be other species belonging to the same genus. In contrast, if 1,000 species go extinct then many genera are likely to die too. Figure 6 depicts this relationship graphically and it can be used to determine how many species go extinct if extinctions at the generic (or family) level are known. Thus, an extinction of 60 per cent genera is shown to require a species extinction of 95 per cent. There is a slight problem with this approach because it assumes that species extinctions are random, and this is unlikely to be the case. Genera are groups of closely related species that usually have similar ecologies. Thus, if a species goes extinct due, say, to some harmful environmental change, then the other species within the same genus are also likely to have been affected. Reports of species extinction rates calculated from generic (or family) datasets, therefore, have to assume that the losses are random rather than focused on ecologically similar groups.

Another issue with calculating extinction rates that Sepkoski had to face was the time intervals that he chose to group his data into. Nowadays these intervals are known as 'time bins'. Geological time is divided into periods, such as the Jurassic and the Cretaceous, but these last too long (their duration spans from twenty-seven to seventy-nine million years in length) to be used as time bins. Periods are subdivided into epochs, then stages, which are in turn divided into zones. There are eighty-four stages (with an average duration of 6.4 million years) and hundreds of zones. Compiling fossil occurrences by zone would potentially produce a high-time-resolution dataset, but many fossiliferous horizons are not dated to a zonal level. Sepkoski's curves therefore show extinctions per stage—a reasonable compromise and probably the best resolution that can be achieved with our current knowledge of the fossil record. Maybe in several hundred years' time, after a lot of palaeontological effort, it will be possible to plot charts showing species extinctions per zone.

Sepkoski's graph clearly showed that extinction rates have varied enormously over time, with the big five mass extinctions standing out as peaks (see Figure 5). Grouping extinction rates of each stage in a ranked order reveals that the mass extinctions are part of a continuum; they do not form a distinct group (Figure 7) or, put another way, as well as the big five there are a lot of lesser extinction crises that gradually grade into the background extinction intensities. This is interesting, but by far the most controversial and extraordinary claim to arise from Sepkoski's original study

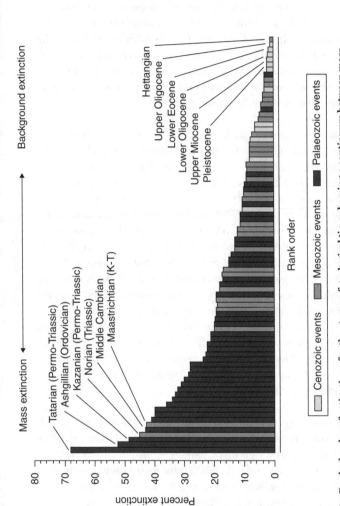

7. Ranked order of extinctions for the stages of geological time showing a continuum between mass extinctions (the end Permian is the greatest value) and background extinctions.

was the suggestion that extinction rates over the past 300 million years show a repeating pattern—a periodicity of twenty six million years. Looking at Figure 5 it can be seen that there is some justification for this idea: extinction peaks in this interval seem to show a reasonably regular spacing, although they certainly differ in magnitude. For example, the peak following the end-Cretaceous mass extinction sixty-six million years ago is rather small.

There is no conceivable Earth-bound mechanism that can produce an extinction periodicity and so a range of extra-terrestrial causes have been suggested, most of which involved periodic bombardments by comets or meteorites. Such ideas were in vogue at the time they were proposed in the mid-1980s because this was only a few years after meteorite impact evidence at the end of the Cretaceous was discovered. A typical suggestion invoked an invisible dark star, called Nemesis, orbiting beyond the outermost planets of the solar system. Every twenty-six million years it disturbed the Oort Cloud—a belt of icy debris in the outer reaches of the solar system, thought to be the source of comets—causing the Earth to be bombarded. Such ideas gained enormous media attention, and it could be argued they have inspired a whole series of Hollywood movies about giant impacts, but they have not generally been accepted in the scientific community—primarily because more recent research has degraded the twenty-six million year pattern: extinction events seem much more randomly spaced than originally thought.

In the 1980s the mass extinction dates were not known with accuracy, but there have since been considerable improvements in the technique of radiometric dating. This uses the fact that unstable atoms like uranium and potassium decay at a known rate which allows the age at which some rocks form (typically volcanic ones) to be calculated. With improvements in measurement precision it is now possible to achieve accurate dates for rocks as much as 300 million years old. For example, in the 1980s the Permo-Triassic (P-Tr) mass extinction was thought to have happened around 248 million years ago, plus or minus several million years either way. Thirty years later, the latest dating estimates suggest this extinction occurred at 251.941 million years ago with an error of less than 40,000 years.

Thanks to this much improved dating of extinction events, their spacing is now known with some accuracy and it is seen to be irregular. For example, a Mid-Permian extinction peak was followed by the end-Permian mass extinction ten million years later. This was followed by an (intra-Triassic) extinction twenty-two million years later, and then an end-Triassic mass extinction thirty million years after that. Clearly, there was nothing regular about the spacing of these extinctions. Also, despite much searching, the end Cretaceous remains the only mass extinction clearly coincident with a major meteorite impact; large-scale volcanism is now the main contender for most other mass extinctions. Nonetheless, some scientists still propose that extinctions, and even volcanism, can be driven by extra-terrestrial causes. A recent idea suggests that the solar system's cycling through the mid-plane of the galaxy every thirty million years causes it to pass through an invisible disc of dark matter. If concentrated in clumps, this material could be captured in the Earth's core causing energy release. Ultimately this may trigger warm plumes to rise through the mantle and erupt as giant-scale volcanism. Imaginative as this notion is, the key problem, as already noted, is the lack of an observed extinction periodicity.

Extinction trends

One of the clearest trends in Sepkoski's graph (Figure 5) is the dramatic decline in extinction rates over the past 560 million years. Thus, background rates in the Cambrian are as high as the mass extinction rates in younger intervals. The reason for this trend is far from clear although there are several ideas available. It may indicate that environmental conditions were much harsher in deep time; or it could suggest the earliest animals were not as hardy as younger ones, and so more prone to go extinct. Around three-quarters of all Cambrian species are trilobites—a group of multi-legged arthropods that scurried around on the seabed. Their evolutionary history was very 'volatile', meaning that new species appeared frequently but they also died off quickly causing extinction rates to be high. Whether this means that trilobites were simply prone to extinction or that they were living in hard times is unclear. Alternatively, elevated early extinction rates may be an artefact of the way that genera are defined by the palaeontologists who name them. A genus is defined as a set of closely related, similar-looking species and it is up to the scientist studying them to decide—often in a rather arbitrary way—when species differences are enough to warrant putting them in another genus. In the Cambrian there were, on average, fewer than three species in each genus but this ratio has generally increased ever since, and today the figure for most animal groups is nearly eight species/genus. Humans are a good example of this, because until recently there used to be more than a dozen species in our genus, *Homo*, though all our close relatives such as the Neanderthals and *Homo erectus* are now extinct.

It may be that palaeontologists studying Cambrian fossils have been over-keen to create genera. The result is that, in the Cambrian, it only requires a few species to go extinct for a genus to go extinct as well, whereas for the more recent fossil record it requires nearly three times as many species to die. So, the elevated early generic extinction rates could be the result of a whole combination of factors (or just one): tough environmental conditions, feeble animals (especially trilobites) prone to extinction, and/or a purely artificial effect caused by palaeontologists assigning relatively low numbers of species to each genus in the Cambrian.

The obverse of the declining extinction rates is a first-order trend of increasing diversity through time (Figure 4)—a clear pattern that was first recognized by palaeontologists as long ago as the mid-19th century. This suggests that origination rates (the rate at which new species appear) have increased and/or extinction rates have declined. The increase has not been constant though, especially in the case of marine animals. Rather there are secondary patterns of plateaus of diversity (especially for the 200-million-year interval from the middle of the Ordovician to the end of the Permian) punctuated by times of rapid rise (Figure 4). The past sixty million years has seen an especially steep diversity increase among marine organisms and the trend is even more spectacularly manifest among terrestrial animals. Whether this is an artefact or a true increase has been much debated. The increase has been attributed to an edge effect of the dataset, called the 'Pull of the Recent'. This idea notes that we have an exceptionally good knowledge of living species compared with the fossil record and so modern diversity is exceptionally high. As a result, species found as fossils are more likely to be recognized as a living species the nearer they are in age to the present day thereby extending species ranges and causing diversity to apparently increase quickly. Conversely, extinction rates decline as the present day is approached.

The significance of the 'Pull of the Recent' has been evaluated by studying bivalve molluscs (i.e. clams such as cockles, mussels, and oysters). This group has a thick shell and they live in marine sediments, both factors that ensure they have a good fossil record—so good in fact that the record of bivalve genera in the lead-up to the present day is 95 per cent complete (and so the 'Pull of the Recent' only accounts for 5 per cent of the observed bivalve diversity increase in the past sixty million years). This, and similar studies of well-preserved groups, indicates that, while the 'Pull of the Recent' undoubtedly occurs, the rapid diversity increase seen in Figure 5 is a real phenomenon—life really has been getting more diverse over the past sixty million years.

Evaluating the patterns shown in extinction and diversity graphs underpins our ability to understand the causes of the observed trends. One of the key debates has been whether the history of global diversity shows a pattern of unconstrained expansion punctuated by mass extinctions, or whether, in-between these crises, diversity is density-dependent. This term means that, as the number of species increases, extinction rates increase due to increasing competition among species as environments fill up. In other words, can environments keep on supporting new species or do they reach a saturation point? If the former is the case, then diversity should increase exponentially (unless an extinction event happens), whereas the latter would produce a logistic growth pattern in which the rate of increase gradually flattens out. The debate between these two styles of diversity increase has been running for many years and has not been resolved mainly because examples of both can be found. Thus, the Mid-Ordovician diversity increase was followed by a diversity plateau, and so looks like a density-dependent trend, whereas the rapid increase of diversity in the past sixty million years looks unconstrained (Figure 4).

The end result of density-dependent diversification is an equilibrium that only allows new groups to thrive at the expense of existing groups. This is the 'one-in one-out' style favoured by Darwin, albeit applied on a larger scale. It is also similar to the idea behind the species–area effect that predicts the numbers of species that an area can support. However, the latter is an ecological model operating on a timescale of years whereas density-dependent diversity predicts trends on evolutionary timescales of millions of years. A closer comparison can be made with the Red Queen hypothesis. This is named after a character from *Alice in Wonderland*, and views evolution as being a continuous process in which species are in constant competition with one another causing them to either constantly evolve or go extinct.

So, what do we find in the history of life? The evolution of mammals over the past sixty million years has provided some informative examples purporting to show density-dependent diversity. One of the most interesting is the Grand Coupure (Great Cut), a major turnover of European mammal faunas thirty-four million years ago during the early Oligocene epoch. This saw the dominant perissodactyls, a group of odd-toed mammals represented by horses today, replaced by mammals that had migrated from Asia, including relatives of rhinos, pigs, and hippos. Oceanic mammals also saw great changes in the early Oligocene: the archaeocetes, whales with long, serpentine bodies, were succeeded by toothed whales and baleen whales, two groups that are still with us today. It has been suggested that these changes are recording the replacement of primitive groups by more advanced groups in environments that were too full to support all of them. However, there is a problem with this idea because coincident environmental changes may have caused the incumbents to go extinct before they were replaced by new forms. Climates cooled rapidly in the early Oligocene and, as polar ice caps started to form, global sea level fell and ocean circulation was transformed. These substantial changes may have been enough to cause the extinction of mammals in both terrestrial and marine settings. Detailed studies of the Grand Coupure turnover in Europe also show that the perissodactyls were extinct before the arrival of the Asian immigrants further supporting an environment-driven cause.

A more convincing example of competition-driven extinction comes from the Great American Interchange. This was caused by the creation of the Panama Isthmus three million years ago which allowed animals to migrate between North and South America. For a short period, the diversity of both continents increased rapidly, but this was followed by a wave of extinctions indicating that these regions could not support such large numbers of species. The result was generally a victory for the North American invaders: about 50 per cent of South American mammal genera hail from the north while only 20 per cent of the North American mammal genera are from the south (e.g. opossums and armadillos). However, the Interchange is an unusual case. Generally, the fossil record of mammals over the past sixty-five million years indicates that environmental changes drive diversity fluctuations and that overall the mammalian radiation has the appearance of being unconstrained.

Extinction selectivity

As well as providing information on extinction patterns and processes, fossils also provide valuable evidence on the nature of selection during extinction. Many attributes can be 'selected', including geographic range, life site, trophic (feeding) group, body size, and reproduction strategy. However, probably the only general 'rule' for extinction is the observation, made for many groups, that broad geographic range is by far the best insurance against extinction. The reason is obvious: a wide distribution ensures that, while local environmental changes can wipe out local populations, a widespread species will have populations elsewhere that survive. This has been shown by studies of extinction rates among bivalve genera. Those that are widespread have lifespans exceeding a hundred million years while more localized (endemic) genera on the whole last <30 million years. Only during mass extinction events does this link weaken, but even at such times it does not entirely disappear. Geographic range is also more important than the degree of specialization of a species (known as its niche breadth). Thus, even highly specialized species can have low extinction risks as long as they occur over wide areas, although on the whole species with a broad range tend to be generalists.

Inevitably species attributes are interlinked, making it somewhat difficult to disentangle their relative importance. For example, many marine invertebrates have planktonic larval stages that drift great distances in ocean currents ensuring their widespread occurrence. Alternatively, other groups have benthic larvae (they live on the seafloor). These are often bigger than planktonic larvae and have higher individual survival rates, but they do not spread as far. Thus, wide geographic range, planktonic larval stages, and low extinction rates often go hand in hand.

Larger organisms are often considered to be at greater risk of extinction, especially at times of mass extinction (and also in the present day). The loss of the dinosaurs at the end of the Cretaceous is a classic example (the birds—which are small, feathered dinosaurs with beaks—survived) and the Ice Age extinctions which removed large mammals like mammoths is another. It is not hard to think of reasons for this: large animals have slow reproduction rates and small populations. However, large animals can also be widely dispersed, a factor which should favour their survivability because of their greater geographic range. These competing traits may account for the unpredictable link between large body size and extinction risk at times of background extinction. Thus, a study of mammals from the Miocene (23.0–5.3 million years ago) of North America showed that *smaller* species had statistically higher extinction rates than large ones, but since that time there has been no size correlation.

Extinction intensity also varies enormously among different groups. Thus, as we have already noted, the bivalve molluscs have very low extinction rates, they are the carthorses of the evolutionary race, with the result that their genera and species are long-lived. In contrast, the racehorses of evolution are the ammonoids—swimming molluscs with coiled shells and a squid-like head—which have much higher rates. In Sepkoski's dataset, bivalve genera show ~10 per cent losses in each stage, whereas the figure for ammonoids is nearer 50 per cent.

It is possible that extinction selectivity has changed through time as new groups arise. This idea lies behind the concept of the Mesozoic Marine Revolution (MMR) proposed by Geerat Vermeij. He noted that many of the main shellfish predators today, such as shell-crushing fish, mantis shrimps, and crabs, first evolved during the Mesozoic Era (252–66 million years ago) thereby making life increasingly dangerous for animals living exposed on the seabed—the epifauna. One strategy to avoid predation is to burrow into sediment and become infaunal. The result has been a very gradual transition of seafloor life since the start of the Mesozoic from epifaunal- to infaunal-dominated communities. Bivalves are one group that clearly show this change. Before the MMR most bivalves lived on the seabed whereas today they are mostly hidden within the sediment (infaunal). The transition has come about because of a slightly higher extinction rate of epifaunal versus infaunal bivalves. Trends such as this are said to be due to predator–prey escalation and are an example of the Red Queen hypothesis in action.

The Paleobiology Database

Most of the discussion of diversity and extinction in this chapter has focused on the database and graphs produced by Jack Sepkoski over a twenty-year period up to the end of the 20th century. Since then several groups of palaeontologists have compiled their own databases of which the best known is the Paleobiology Database (www.paleobiodb.org)—known as the PBDB. This differs in several regards from Sepkoski's, most notably because it groups data into time intervals that are nearly twice as long (eleven million years) and it also includes information on occurrences. The lower temporal resolution is unfortunate because it makes it impossible to distinguish closely spaced extinction events, they just get lumped together, but it has been chosen because the age assignments of many fossils is imprecisely known. The inclusion of occurrence data is an improvement though because it allows biases such as variable sampling intensity to be investigated. John Alroy has pioneered these analyses and he especially favours the use of subsampling. This technique aims to avoid the problems of uneven sampling of the fossil record. For example, if a particular time bin has been well-sampled (lots of fossil collecting) then diversity will be artificially higher than poorly sampled intervals. Subsampling uses a computer program to randomly select samples of fossil occurrences from the larger datasets and equalizes their size to that of a smaller sample. This approach has an especially notable effect by reducing the apparently high diversity seen in the past few million years. Rocks of this age are widespread and they have been better sampled than older examples with the result that diversity levels seem high.

So what does analysis of the PBDB reveal? In essence it shows that the major phenomena identified in Sepkoski's graphs are still apparent which is not that surprising given the first-order, large-scale nature of the patterns he found. The PBDB studies confirm the high Cambrian extinction rates followed by a long-term decline and the presence of mass extinctions. Only the post-Cretaceous diversity rise is a little less spectacular than originally seen.

Chapter 4
The great catastrophes

What is a mass extinction?

The idea that mass extinctions are a feature of life is well-established as is their intimate link to cataclysmic events such as giant meteorite impacts and huge volcanism. Their recognition and acceptance as genuine phenomena over the past forty years has been responsible for the rebirth of catastrophism in scientific thought, after over 150 years of being out of favour. Despite their popularity, the term 'mass extinction' is often applied a little too widely. Part of the problem can be seen in Figure 7 which shows there is a continuous gradation from mass extinctions into background extinctions, which raises the question of where to put the cut-off? The best way is to consider the attributes of the five largest crises which were unequivocally mass extinctions.

Mass extinction events are geologically short intervals of time (always less than a million years), marked by dramatic increases of extinction rates in a broad range of environments around the world. In essence they are global catastrophes that left no environment unaffected. They are also seen as clear peaks in extinction rate plots such as those of Sepkoski (Figure 5). The global scale of mass extinctions is clearly illustrated by the end-Cretaceous event that saw dinosaurs eliminated from landmasses that ranged from the poles to the equator, while the oceans saw the simultaneous disappearance of many planktonic species and many other groups besides. Indeed, major losses among diverse planktonic groups are a distinctive feature of the big five mass extinctions; surface-water catastrophes are not seen during background intervals. Such wide-ranging extinctions clearly require global-scale causes, and they contrast with more environment-specific and regional crises caused by normal, background processes such as interspecific competition and habitat loss. This does not mean that normal processes are irrelevant though. Several studies have shown that, for some mass extinctions at least, the losses show the same patterns as those encountered during background times. Put another way, groups with the highest extinction rates during background intervals often have the highest proportional losses during mass extinction events. This was especially the case for during the P-Tr mass extinction when groups that had high background extinction rates, such as ammonoids and brachiopods (a shellfish group that lives inside two valves attached to the seafloor), suffered enormous losses, while groups with low background rates, such as the bivalves, had relatively modest losses.

During background intervals, broad geographic range is by far the best insurance against extinction because a widespread species facing a local crisis will survive elsewhere thanks to its widely distributed populations. For a global crisis, species no longer have a refuge and so, potentially, different factors may become important for survival. Much research is therefore expended on investigating the attributes that allow species to survive mass extinctions because it can help us understand what caused them. The end-Cretaceous mass extinction provides a nice example. During this terrible crisis, cold-blooded groups such as snakes, crocodiles, and turtles had proportionally fewer extinction losses than warm-blooded groups like the dinosaurs, birds, and mammals, suggesting that lower metabolic rates were an advantage. Equally informative, during the P-Tr mass extinction marine invertebrate groups best able to tolerate low oxygen levels, high temperatures, and lowered pH (more acidic conditions) seem to have fared better, pointing to factors that may have been important at this time.

By identifying extinction selectivity, it becomes possible to speculate on the nature of the environmental changes responsible for extinctions. Nonetheless, despite the severity of mass extinctions, geographic range still appears to confer some survival advantage. This is shown by the fate of North American mammals during the end-Cretaceous crisis. Recent work has revealed that only four species survived the extinction (from a total of fifty-nine species) and these were all among the most abundant and widest-ranging animals. Care has to be taken though, when evaluating the success of survivors, to ensure that the correct trait is being given the credit. For example, it has been argued that enormous losses of thick-shelled fossils during the P-Tr mass extinction was due to the difficulty of secreting shells in acidified oceans. However, low-oxygen conditions are also inimical to thick-shelled species and so this factor could have been the significant one.

The immediate aftermath of mass extinction is often marked by the dominance of a few species that are both abundant and widespread. These are known as *disaster taxa* and their success stems from the fact that they have little competition in a world denuded of many species. Examples include the bivalve *Claraia*, which became pandemic (globally distributed) after the P-Tr mass extinction, and the brief proliferation of ferns after the end-Cretaceous event. This latter bloom is known as the 'fern spike' because fern spores become abundant in collections of pollen and spore grains at this time. Another interesting feature of the post-mass extinction interval is that the survivors are often much smaller than those present beforehand. This has been called the *Lilliput effect*, after the land inhabited by tiny people in Jonathan Swift's 18th-century novel *Gulliver's Travels*. There are several possible causes for this, including the preferential loss of large species during the extinction event (leaving just small survivors), the reduction in size of species as they cross the extinction boundary, or the evolution of new, small species during the immediate aftermath.

Mass extinctions have fundamentally changed the trajectory of life often by removing the dominant species (known as the incumbents) from environments. The replacement of dinosaurs by mammals sixty-six million years ago is the classic example of this phenomenon. Ironically, the success of dinosaurs was also due to the loss of the previous incumbents during an earlier mass extinction just over 200 million years ago. This end-Triassic event eliminated numerous animals leaving the dinosaurs to radiate and enjoy 135 million years of dominance. Interestingly, the mammals also survived the end-Triassic mass extinction but they had to wait for another mass extinction before they could become the dominant animals on land.

Mass extinctions also cause major changes in the evolutionary history of groups that survive because they filter out many types. This is seen among trilobites during the end-Ordovician mass extinction. Prior to this crisis trilobites had pursued a broad range of lifestyles ranging from species that lived in the upper water column, where large forms swam and tiny forms drifted, to ones that lived in the deep sea and burrowed in mud. The severe losses among trilobites at the end of the Ordovician included all the water-column-dwelling species and the group never again returned to this habitat. Some groups can survive but fail to recover from their mass extinction losses and instead just persist at low diversities, eventually fading away and going extinct. These have been vividly named *zombie taxa*, the *bellerophontids* provide a good example. These were a successful group of snails before the P-Tr mass extinction but they suffered major losses leaving only a few species in the seas of the Early Triassic. These plodded on, zombie-like, for a few million years before they faded away, leaving no descendants.

As well as Lilliputians and zombies, the aftermath of mass extinctions is marked by *Lazarus taxa*. These get their name from a biblical character who was brought back to life by Jesus. Lazarus therefore had a period of time when he was dead, and this attribute also seems to happen to some survivors of mass extinctions; they disappear during the crisis but then reappear, many millions of years later, as if risen from the dead. Obviously, it is impossible for an extinct species to come back to life but the Lazarus effect is clearly telling us something interesting about the post-extinction interval. One possibility is that the species are confined to some local refuge, undiscovered by palaeontologists, and only reappear once they become widespread. Another reason could be that some species become very rare in the extinction aftermath, possibly because conditions were harsh, making it highly unlikely that their fossils will ever be found. Only when conditions ameliorate do they become more abundant and turn up again in the fossil record. Another possibility, is that the reappeared species are actually newly evolved forms that closely resemble previously extinct species. This third alternative may be the most important, because the fossil record is full of examples of such convergent evolution. It is clearly seen in the aftermath of the P-Tr crisis when many species of sponges and calcareous algae disappeared only to apparently reappear tens of millions of years later. These organisms have such simple morphologies that it probably did not take evolution much effort to reinvent their form.

The first extinctions

For the first few billion years of Earth history, during the Precambrian Eon, evolution was on the microbial, single-celled level. This changed around 540 million years ago when animals started to diversify quickly in a time known as the 'Cambrian explosion'. However, the thirty million years before this saw the appearance of some unique, enigmatic fossils known as the Ediacaran fauna. Their abrupt disappearance at the base of the

Cambrian raises the possibility that Earth suffered its first mass extinction at this time.

Ediacarans were broad, flat organisms that came in a range of shapes including discs, elongate ovals, and frond- or feather-like forms built of repeating segments. They are generally preserved as sandstone casts and were likely entirely made of soft tissue. With few exceptions they do not appear to have been able to move and instead either lay flat on the seabed or raised themselves above it while being attached to a stalk and holdfast. The Ediacarans have given palaeontologists plenty to ponder. Some regard them as a unique grade of organism unrelated to anything living, while others assign them to extant groups such as worms and soft corals. If the former view is correct then their disappearance at the end of the Precambrian represents a major extinction. Environmentally, there was certainly a lot happening at this time, especially to levels of ocean ventilation, to suggest a possible cause for such an event. The extinction could also have been driven by the development of actively burrowing organisms that disturbed the sediment on which the Ediacarans sat. However, without any clear idea of what the Ediacarans were, their demise is currently unclear.

With Ediacarans out of the way at the start of the Cambrian, the fossil record rapidly became dominated by the trilobites, an arthropod group, while the first reefs appeared in warm, shallow waters. These were made of sponges, belonging to a group called the archaeocyathids. Sites of exceptionally good fossilization quality such as the Burgess Shale in British Columbia, Canada, and in Kunming, China (the Chengjiang biota), also contain a lot of additional soft-bodied animals that show that Cambrian oceans were teeming with diverse life. However, the trilobites have the best fossil record, thanks to their readily fossilized calcite skeleton, and much of what we know about early extinction events is based on the fortunes of this group. Their early history certainly seems to be a roller-coaster ride of extinctions and recoveries. This resulted in some of the highest extinction rates of the fossil record, and several

Cambrian intervals are regarded as mass extinctions. The most severe was around the end of the Early Cambrian, when the archaeocyathid reefs disappeared and, a short time later, two of the dominant trilobite groups—the redlichiids and olenellids—also went. There are further trilobite extinction events in the later Cambrian and they mark the boundaries of biomeres, which were intervals of time characterized by distinct trilobite populations.

The causes of all the Cambrian crises is poorly known, mainly because they have received little study. However, most of them coincide with sea-level changes and frequently the losses occurred during sea-level rise (transgression) when oxygen-poor (anoxic) waters become widespread. The end of the Early Cambrian extinction also seems to coincide with the eruption of huge volumes of flood basalts in Australia (known as the Kalkarindji Province). This is the first of many links between extinction episodes and volcanism, and the cause-and-effect relationships may be similar to those encountered during later crises, but at the moment our understanding of Cambrian environmental conditions is best described as under-studied.

End-Ordovician mass extinction

Marine diversity increased considerably in the Ordovician and the trilobites were joined by numerous other groups including the brachiopods, the graptolites (a group of organisms that drifted around the oceans in little, stick-like colonies), and the conodonts (primitive, eel-like fish with tiny, spikey teeth that fossilize well). Following the loss of the archaeocyathid sponge reefs in the Cambrian, reefs had reappeared in the Ordovician and this time they were constructed by corals, belonging to a group called the tabulates, and a heavily calcified group of sponges called the stromatoporoids. All of these groups were badly affected by a crisis in the Hirnantian (the final stage of the Ordovician) around 445 million years ago. This is widely regarded as the first true mass extinction and is the first of the big five (Figure 5).

Losses were exceptionally severe among animals living in the surface waters. As noted above, all swimming and free-drifting trilobites disappeared along with all but a handful of conodont and graptolite species. Even those trilobites that had a planktonic larval stage but a benthic adult stage suffered—thus, the benthic trinucleids were abundant before the mass extinction but their possession of an open-ocean larval stage seems to have been a factor in their extinction. The take-home message from these losses is that environmental change in the surface waters must have been exceptionally severe at the start of the Hirnantian. Seafloor life did not escape though. The bottom-living brachiopods were also badly affected with recent estimates suggesting 85 per cent of species vanished. The total extinction at the level of genera among all groups was around 45–50 per cent making the end-Ordovician event second only to the P-Tr mass extinction in magnitude (Figure 5).

The end-Ordovician mass extinction seems to have occurred in two distinct episodes. The first and most serious occurred at the start of the Hirnantian Stage, when most groups in the water column died out. There was then an interval of nearly half a million years before a second extinction pulse that eliminated further brachiopods and trilobites. The intervening interval is characterized by the development of the *Hirnantia* fauna: a low-diversity group of cool-water species (named after a brachiopod) which became very widespread at this time. The expansion of the *Hirnantia* fauna around the world coincided with a major glaciation episode in the southern hemisphere.

During the Late Ordovician, North Africa lay over the South Pole and it was here that a large ice sheet grew during the Hirnantian. The first phase of the end-Ordovician mass extinction thus coincides with a glaciation and also a major sea-level fall caused by the trapping of water in continental ice sheets. The melting of these ice sheets, half a million years later, caused the sea level to rise again and also saw the spread of anoxic waters in deep water that encroached into shelf seas. These substantial environmental changes coincide closely with the mass extinction phases. In addition, recent studies have provided indirect evidence, through elevated mercury concentrations in sedimentary strata, for major volcanism at this time providing a potential culprit for the crisis. Mercury is a volatile metal that is erupted with volcanic gases, and is ultimately removed from the atmosphere and buried in sedimentary rocks. The issue for the end-Ordovician volcanism is that the site of the eruptions is not yet known.

Late Devonian mass extinction

The second of the big five mass extinctions, the Late Devonian event, occurred 374 million years ago near the boundary between the Frasnian and Famennian stages, consequently it is often known as the F-F mass extinction. Generic extinction levels were around 35 per cent (although estimates vary from as high as 60 per cent to as low as 20 per cent) making it the least severe of the big five. The victims once again included the brachiopods, the trilobites, and most reef life. The inhabitants of the water column were also devastated. With the exception of fish, most pelagic groups (a term that includes both swimming and planktonic forms) of the Late Devonian were badly affected. These included the ammonoids (distant relatives of squid that inhabited spiral shells), conodonts, and the cricoconarids (an enigmatic group of tiny conical shells thought to have had a planktonic lifestyle) which disappeared entirely. Some have argued that the F-F mass extinction was more to do with a decline in the rate at which new species were appearing (the origination rate) rather than an increase in extinction rates, and have termed it a 'mass depletion event'. Marine diversity certainly seems to have been in long-term decline for several million years before the F-F boundary. However, detailed collecting from rock successions that straddle the F-F boundary shows that many extinction losses occurred in a very limited amount of strata at the end of the Frasnian Stage indicating that there was a clear-cut, abrupt mass extinction at this time, even if it did come at the end of a period of decline.

Mass extinctions, by definition, should affect life in all habitats and, by the later Devonian, life had successfully colonized land, especially in wet areas and river bank locations where amphibians were making the first footprints. The Frasnian Stage also saw the spread of the first forests which were dominated by a tree called *Archaeopteris*, but the heyday of this terrestrial community was short-lived because it did not survive beyond the Frasnian. This suggests that the F-F crisis was the first to affect terrestrial communities. It is difficult to judge the contemporaneous fate of the amphibians, because their fossil record is very sparse, but it is possible that they too suffered a major extinction at this time.

The cause of the F-F crisis is not well-understood. In many marine sections the boundary is marked by two pulses of anoxic deposition called the Kellwasser Events after two organic-rich limestones in Germany. Sea-level oscillations also seem to have been substantial at this time and they have also been implicated in the crisis. However, there are many anoxic events and sea-level changes in the Devonian and they do not coincide with mass extinctions, so why should these examples be more lethal to marine life? It is possible that the F-F changes were much more intense. Unlike earlier anoxic episodes in the Devonian which were restricted to deeper waters, the Kellwasser anoxia developed over a broad range or water depths from deep shelf settings into very shallow waters. The latter development is unusual, and the spread of anoxia would have severely curtailed the area of shallow marine habitat during this crisis. Others have suggested that extinction losses were concentrated among warm-water species thereby indicating that cooling may have caused the crisis. Chemical proxies for temperature change certainly support the idea that there was a decline of ocean temperatures during the F-F interval.

Finally, recent study of F-F boundary sediments has revealed enrichment in trace concentrations of mercury that likely records a major phase of volcanism at this time. As with the end-Ordovician volcanism, the site of major eruptions during F-F times is not yet known. Suffice to say that the cause of the Late Devonian mass extinction is still under discussion; there are many of the usual suspects that appear during later extinction events, but for this older example some aspects are proving elusive.

End-Devonian extinction

The end-Devonian crisis, around 360 million years ago, is often known as the Hangenberg event, because of an organic-rich limestone of this name that developed at the Devonian-Carboniferous boundary in Germany. It is best known for the devastating impact it had on fish, the worst in their entire history. This group had diversified tremendously during the preceding Devonian (the period is often known as the 'Age of Fish') but many of them failed to survive into the Carboniferous. The victims included giant arthrodire placoderms—an impressive group of predators with armoured skulls and slicing, scissor-like jaws (Figure 8)—and marine lungfishes—a group that is still with us today but now restricted to freshwater environments. Trilobites, frequent victims of mass extinction, also suffered major losses during the Hangenberg crisis and this time they never really recovered (they became a zombie group, quietly living out their days for another hundred million years without doing anything spectacular). It also seems likely that the amphibians—still taking their first tentative steps on land—suffered extinctions, pointing to a crisis on the land as well as in the sea.

There was no shortage of major environmental changes during the Devonian-Carboniferous interval that may have caused the extinction. Late in the Famennian (the final stage of the Devonian) a major sea-level rise saw the spread of anoxic waters in many shelf seas. Many of the marine extinction losses seem to

8. **Examples of Late Devonian marine fish that failed to survive the end-Devonian mass extinction.** *Dunkleostus* (top), a 6-metre-long arthrodire, *Pteraspis* (bottom right), a 20-centimetre-long heterostracan, and *Lepadolepis* a 1-metre-long placoderm.

coincide with the development of these inimical conditions. This was followed by cooling, which culminated in the growth of ice caps on southern hemisphere continents, and a global sea-level drop of around 100 metres. A secondary extinction event occurred around this time, with shallow marine invertebrates being the main victims. Interestingly, this series of events—ocean anoxia followed by cooling and sea-level fall—is the opposite order to that seen during the end-Ordovician mass extinction, when glaciation preceded anoxia. This suggests that rapid, large-scale environmental changes are responsible for the extinctions, but the order in which they occur does not seem to matter. What is singularly lacking from the end-Devonian extinction story is any evidence of contemporaneous, large-scale volcanism, or at least no evidence has been found yet.

A Mid-Carboniferous enigma

The recovery from the end-Devonian extinction was remarkably slow. Reefs were devastated by the F-F crisis and the Hangenberg event was their coup de grâce; they are missing from the seas and oceans for much of the succeeding Early Carboniferous Period. The first small, simple reefs, made of rugose corals and chaetetids (another, heavily built group of sponges) did not appear for over twenty million years. This is the longest absence from the fossil record in the entire history of reefs, and it is mirrored by a rather muted recovery among marine invertebrates. Eventually though, things started to get better, and in the later part of the Early Carboniferous (specifically in the time interval known as the Asbian Stage) many new species started to appear. These included forams (protists that secrete a chambered shell that is typically less than a millimetre in size) and many types of brachiopods, especially the gigantoproductids which reached sizes of up to 30 centimetres (which is large for a brachiopod). The recovery coincides with a warming trend, and the two phenomena may be related because diversity generally correlates positively with temperature. For example, the most diverse habitats today are in equatorial settings.

All was going well for life in the middle of the Carboniferous but then something seemed to go wrong, and diversity began to decline, especially among the dominant marine groups such as the rugose corals, brachiopods, and crinoids (stalked organisms with a head of feathery arms—they can still be found in modern oceans where they are known as feather stars and sea lilies). This occurred during the Serpukhovian Stage but whether this interval deserves to be recognized as a time of true mass extinction is difficult to judge—it certainly does not stand out in extinction charts like Figure 5. However, it may be that origination rather than extinction was the factor here. Diversity is controlled by both the rate at which species go extinct and the rate at which new ones appear. For the Serpukhovian crisis it appears the decline was caused by the failure to replace losses with new species.

The marine diversity decline in the Mid-Carboniferous is closely tied to an intensification of glaciation and global cooling. Continental ice sheets probably existed throughout much of the Carboniferous, especially in southern polar latitudes, but they became much more extensive during the Serpukhovian. Far away from the ice, equatorial waters also seemed to have cooled causing the loss of many warmth-loving marine invertebrates. The trapping of water in ice sheets also caused sea level to fall, and thereby it saw the area of shallow sea habitats shrink as the waters retreated to the continental margins. Curiously though, there was no associated Serpukhovian crisis among the extensive equatorial forests, where the inhabitants (amphibians and primitive reptiles) showed no major changes.

So does the Serpukhovian diversity decline constitute an extinction event? Some palaeontologists think so and have promoted it into the ranks of the 'big five', but it lacks the attributes of the true mass extinctions. It was not a global crisis but rather was concentrated among equatorial marine invertebrates, barely manifest on land at all. The speed of the diversity decline is also unclear, the big five mass extinctions all happened quickly (in a geological sense) whereas the Mid-Carboniferous diversity fall may have been slower and lasted longer. Nonetheless, the Serpukhovian crisis provides a nice example of how glaciation and cooling can reduce diversity levels in marine habitats.

Mid-Permian (Capitanian) extinction

This is the newest of the extinction crises; new in the sense that it was only discovered in 1994, because it actually happened around 262 million years ago during the Capitanian Stage of the Mid-Permian. It is also known as the Guadalupian extinction—an alternative name for the Mid-Permian. The Capitanian is the final stage of this interval and the extinction happened within the middle of it. This crisis occurred about ten million years before the huge P-Tr mass extinction, and for a long time the extinction losses of the older event were included in the terminal Permian extinction. Estimates of generic extinction rates in the Capitanian Stage are around 35 per cent, almost making the extinction of comparable magnitude to the big five. Like other mass extinctions, the crisis affected life on both land and in the sea. The marine extinctions were severe among brachiopods, ammonoids, and forams, while the alatoconchids (a bizarre group of giant, tropical bivalves) were wiped out.

On land the dominant dynasty of animals—the dinocephalians—also disappeared late in the Capitanian. These were a successful group of often very large animals, both predators and herbivores, that were often characterized by their bizarre knobbly heads (Figure 9). They belonged to the therapsids, a broad category that includes many groups of four-legged animals (tetrapods), including the mammals. Following their demise, the dinocephalians were replaced by diverse tetrapod communities that included large reptilian herbivores (pareiasaurs), gorgonopsids (large predators), and smaller herbivores with two pairs of tusks but no teeth (dicynodonts) (Figure 10). This new assemblage had a relatively short lifespan because the P-Tr mass extinction eliminated most of them.

The Capitanian crisis coincides with the onset of a major series of eruptions that covered a tropical seaway in south-west China with vast outpourings known as flood basalt lavas. These are called the Emeishan Traps, and their coincidence with the marine extinctions in the region clearly indicates that volcanism played a role although the exact mechanism is unclear. Some have suggested an episode of global cooling (based on the observation

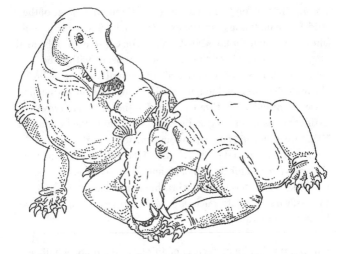

9. **Examples of dinocephalians, the dominant group of terrestrial tetrapods, that were wiped out by the Mid-Permian mass extinction. A sleeping _Estemmenosuchus_, a 3-metre-long omnivore, is about to be woken up by an _Anteosaurus_, a 5-metre-long carnivore.**

that shallow-water tropical species were particularly badly affected) while others have pointed to the development of oxygen-poor conditions in the oceans. Currently there is no clear consensus. Much of the debate about how giant volcanism can cause mass extinction has focused on the nature of the subsequent mass extinction, at the end of the Permian.

Permo-Triassic mass extinction

The third and largest of the big five mass extinctions happened around 252 million years ago and is usually known as the end-Permian or P-Tr mass extinction. The crisis marks the biggest turnover of the fossil record and, although at one time it was thought to have been spread over the final few million years of the Permian, recent study shows the losses were concentrated in perhaps as little as 100,000 years straddling the P-Tr boundary.

10. Examples of typical Late Permian tetrapods that replaced the dinocephalians following the Mid-Permian extinction. From top downwards, *Inostrancevia* (a predatory gorgonopsid up to 3.5 metres in length), *Scutasaurus* (a 3-metre-long, herbivorous paraeiosaur), and *Fortunodon* (a 1-metre-long dicynodont herbivore).

Generic-level extinction losses were around 70 per cent (with species extinctions estimated at up to 95 per cent) and were equally severe on land and in the sea.

The P-Tr mass extinction was the only crisis to have completely eradicated entire terrestrial ecosystems—many plants disappeared along with animal communities. The aftermath was marked by a bizarre world, with only shrub-sized vegetation, that lasted for several million years. Late Permian forests were dominated by seed-producing plants called gymnosperms, a group which includes conifers, and their substantial extinction losses include the glossopterids (a pteridosperm or seed fern group) that thrived in high southern latitudes and an advanced group called the gigantopterids that lived in the tropics.

The fossil record of spores and pollen at the P-Tr extinction level shows several intriguing aspects: mutated pollen (badly formed pollen grains with malformed air sacs), spore tetrads (spores are formed in clusters of four but they normally break up into individual spores when dispersed on the wind, in the case of the P-Tr examples the separation often failed and they stayed stuck together as tetrads), and abundant fungal spores (such spores are typically rare but at the P-Tr boundary they briefly become prolifically abundant). These are all important clues to the damaged and stressed nature of terrestrial plant life. The fungal peak suggests that processes of plant decay which today is achieved by a combination of insect herbivory and fungal attack were not operating in their usual way.

The P-Tr mass extinction was equally devastating in the oceans. The radiolarians are a planktonic group of protists that secrete beautiful skeletons of silica around 0.1 millimetres in diameter. They have an excellent fossil record that goes back 540 million years, but they almost disappeared at the end of the Permian and their recovery only began two million years later, indicating there was clearly something amiss in the ocean's surface waters for a long time. On the seabed the extinctions affected every group with trilobites, corals, and most echinoderms disappearing along with many molluscs. Brachiopods, which had been one of the most abundant and diverse marine groups during the previous 230 million years, suffered more than 90 per cent species extinction. For example, in the shallow seas that covered South China, of the eighty-five brachiopod species present immediately before the extinction only one was present afterwards. Animals with soft bodies, such as worms, generally do not fossilize, but their presence (and diversity) is indirectly recorded by the various burrows they make. These are known as trace fossils, and the disappearance of many types during the P-Tr mass extinction indicates that the crisis was also severe among the soft, non-shelly animals.

The world in the aftermath of the P-Tr mass extinction was a strange place. Diversity was very low but the few survivors, the disaster taxa, were often abundant and incredibly widespread (they had the world to themselves, after all). In the ocean, a few bivalves were dominant, especially species belonging to the genus *Claraia*, while on land a dicynodont called *Lystrosaurus* become one of the most widespread and abundant terrestrial animals of all time (Figure 11). The dicynodonts would go on to be one of the most abundant herbivore groups of the Triassic.

Another extraordinary facet of the P-Tr crisis was the sheer length of time it took to recover. Early Triassic communities were generally of very low diversity and it was only in the Mid-Triassic, five million years after the mass extinction, that appreciable diversity was re-established in marine ecosystems. This extended interval of recovery is around an order of magnitude longer than

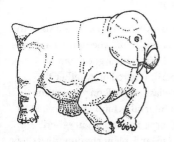

11. **The dominant animals in the ocean and on land in the immediate aftermath of the P-Tr mass extinction: the bivalve *Claraia* (up to 5 centimetres in height) and the dicynodont *Lystrosaurus*, which were around 2 metres in length.**

that seen after other mass extinctions, although the Early Carboniferous was also rather slow but the levels of diversity were not as low as seen in the Early Triassic. The failure to find much of a recovery is generally though to reflect the prolongation of harsh conditions in the Early Triassic. Indeed, conditions became so bad that another extinction occurred within the Early Triassic that wiped out many species of ammonoid. The delayed recovery is primarily a feature of the marine realm; terrestrial communities showed a much quicker recovery time. For example, the temnospondyl amphibians, a group of crocodile-like, large animals, diversified rapidly in high latitudes during the Early Triassic.

Nearly all studies of the P-Tr mass extinction lay the blame at the door of the Siberian Traps—a vast province of flood basalts that erupted in a brief interval at this time. Most attention is focused on the effects of the associated gas emissions, especially CO_2, sulphur dioxide, and halogens. The first seems responsible for the associated rapid warming trend that occurred during the extinction, and the last may have impacted atmospheric ozone generation and thus caused the terrestrial plant losses. This volcanism-extinction connection is explored more fully in Chapter 5.

End-Triassic mass extinction

Despite being one of the big five mass extinctions, the status of the end-Triassic event (201 million years ago) only became established in the 1980s, and some palaeontologists still argue about its magnitude. The debate has primarily centred on whether the losses occurred gradually over the last few million years of the Triassic, or in a geologically brief interval (<1 million years). Most recent studies favour the latter alternative, and claim generic losses of more than 40 per cent. On land, plant communities show substantial changes, although with only modest actual extinction losses, whereas animal extinctions were much more severe. Before the extinction, terrestrial communities were composed of diverse

12. **Examples of Late Triassic tetrapods. Clockwise, from top left:**
Plateosaurus (a 10-metre-long herbivorous dinosaur); *Pseudopalatus*
(a semi-aquatic, carnivorous phytosaur up to 5 metres long);
Desmatosuchus (a 4-metre-long herbivorous aetosaur); *Lophostropheus*
(a bipedal, carnivorous dinosaur up to 3 metres long and a mass
extinction survivor); *Rauisuchus* (a carnivorous rauisuchian, up to
4 metres in length).

groups (Figure 12) such as the dinosaurs (represented by the large,
herbivorous sauropods, and small carnivores) and numerous types
of crurotarsan (a diverse reptile group that included the large,
predatory, quadrupedal rauisuchids, the heavily armoured,
herbivorous aetosaurs, and the large, crocodile-like phytosaurs).
Nearly all crurotarsans disappeared at the end of the Triassic and
their few survivors later gave rise to the crocodilians. Dinosaur
extinctions were less severe, and their rapid recovery marked the
start of their 135 million year hegemony as the dominant

terrestrial animals. In the oceans the end-Triassic mass extinction saw reefs wiped out (yet again), along with all their constituent species of corals and sponges, many species of bivalve, the last conodonts, and nearly all ammonoids. Fortunately, a few species of the last group survived and they rapidly diversified to fill Jurassic seas with their beautiful spiral shells.

Like other mass extinctions, the end-Triassic event precisely coincides with the eruption of huge amounts of lava. In this case it is the Central Atlantic Magmatic Province (CAMP) and the lava flows can be seen today in Morocco, Brazil, and North America. Coincident environmental changes were also similar to those seen during the other flood basalt eruptions, notably rapid global warming. However, unlike most other extinctions, marine anoxia does not appear to have played a role in the end-Triassic event; rather, oxygen-poor conditions were widespread only after the crisis. Instead, ocean acidification (caused by emission of CO_2 from CAMP) is the most popular kill mechanism for this mass extinction while intense warming has also been mooted as the cause of the terrestrial losses.

Early Jurassic (Toarcian) extinction

Twenty million years after the end-Triassic mass extinction, during the Toarcian Stage in the Early Jurassic, the world experienced the eruption of another extensive series of basalts. Today these flows have become separated by continental drift and they are found in the Karoo region of South Africa and the Ferrar region of Antarctica. The Karoo-Ferrar eruptions coincided with a series of climatic and oceanic changes that are closely similar to those seen during the eruption of the Siberian Traps at the P-Tr boundary. Thus, the oceans warmed up rapidly, sea level rose, and oxygen-poor conditions became widespread both in the open ocean and in shelf seas. Despite these similarities, the contemporaneous extinction event was only a mini version of the P-Tr mass extinction (25 per cent generic extinction compared with 70 per cent). No major groups disappeared from the sea and there was no terrestrial crisis. Just why the Toarcian extinction was so subdued, despite the many similarities, is unclear. Nonetheless extinctions occurred among the ammonites, bivalves, forams, brachiopods, and ostracods (tiny crustaceans that live within two valves) with the most severe losses among the last two groups. This crisis closely coincided with the spread of anoxic waters.

Cretaceous-Paleogene mass extinction

The Toarcian crisis was the last major extinction event of the fossil record and the succeeding 180 million years have been pretty much plain sailing for life on Earth (Figure 5), with the single, obvious exception of the most famous mass extinction of all. This occurred sixty-six million years ago at the end of the Cretaceous and was originally called the K-T event, where K is for *Kreide*, German for the Cretaceous, and T is for the Tertiary Period. It is now more commonly referred to as the K-Pg event where Pg stands for the Paleogene Period.

Around 40 per cent of all genera disappeared during the K-Pg event, with the extinction being at its most severe among large animals. The dinosaurs were the best known victims with *Tyrannosaurus*, the horned ceratopsians, and the duck-billed hadrosaurs being among the many losses. The pterosaurs (giant flying reptiles) also succumbed but the birds survived despite substantial losses. Birds are a group that sits within the dinosaur evolutionary tree, and so strictly speaking the dinosaurs did survive the K-Pg event. Therefore, it is more accurate (and pedantic) to say that only the non-avian dinosaurs went extinct. Mammals survived the crisis, but suffered huge losses, as did many freshwater animals (crocodiles, turtles, and fishes).

The abruptness of these terrestrial extinctions has long been debated. The prevailing view in the 1970s was that dinosaurs were in terminal decline long before the K-Pg boundary so that, by the end, there were only a few species left. Intensive collecting efforts have since shown that, on the contrary, dinosaurs were still diverse immediately before the boundary. Many (but not all) palaeontologists now consider the extinction to have been abrupt. This debate is an important one because the speed of extinction helps distinguish between different causes: a giant meteorite impact would produce an instantaneous catastrophe whereas more protracted changes like climate change would not. Plant records also show an abrupt change at the K-Pg boundary, especially among the pollen and spores of North America, although there are no major extinction losses.

Marine extinctions also provide further clues as to the nature of the K-Pg crisis. Losses were terrible among the dominant pelagic groups: marine reptiles, ammonites, and the squid-like belemnites all disappeared. Extinctions among fishes and sharks were also substantial, especially among large-bodied groups. The nautiloids were the only cephalopods with an external shell to survive the extinction and they remain, albeit as a minor component of marine life today. On the seafloor losses were also severe: a detailed study of fossils from Antarctica showed that of twenty-nine species of bivalve and gastropod present immediately before the K-Pg boundary, eleven survived into the Paleogene. The most detailed marine extinction story comes from calcareous plankton thanks to their abundance in sedimentary rocks. Thus, planktonic forams show the abrupt loss of between 70 and 80 per cent of their species at the K-Pg boundary. The immediate survivors are small, simple, widespread forms that were probably ecological generalists.

The substantial K-Pg losses coincide with the major basalt eruptions in western India: the Deccan Traps (Figure 13). Much effort has been expended on dating this volcanic pile, and we now know that volcanism began around 300,000 years before the K-Pg boundary and reached its climax pretty much at the same time as the mass extinction, before quietening down in the Early

13. Example of the stacked lava flows of the Deccan Traps, producing a distinct, stepped topography.

Paleogene. If this was the only thing happening at the end of the Cretaceous then the crisis would be just another example of an extinction-volcanism connection. But of course a giant meteorite also struck the Yucatán Peninsula in Mexico, producing the Chicxulub crater, at the precise moment of the extinction, and the past forty years has seen a seemingly endless debate (summarized in Chapter 5) as to which cause, impact or volcanism, was primarily responsible for the mass extinction.

Chapter 5
How to kill nearly everything

This chapter focuses on the various proposed causes of mass extinctions, often called the kill mechanisms. Over the years an almost bewildering array of ideas have been put forward as likely bringers of death. Many of these are fanciful or lack much, if any, supporting evidence, but most of the debates described here have focused on just a few culprits, notably volcanism and meteorite impact, because their timing is closely coincidental with the extinctions. However, it is important to note that while such culprits may be the *ultimate* cause of a crisis, it is their consequences that likely lead to *proximate* (or direct) causes of extinction. Thus, volcanic eruptions in Siberia are widely held as the ultimate cause of the P-Tr mass extinction, because they occurred at the same time, but the lavas themselves had no direct effect on most organisms. Instead it is the cascade of environmental changes, triggered by the associated volcanic gas emissions, that are likely to be the proximate cause of this great crisis, and understanding these changes is not easy.

Large igneous provinces

Volcanic eruptions are inextricably linked with images of volcanoes but these are in fact among the smallest manifestations of volcanic activity on Earth and do not really constitute a danger to life except in their immediate vicinity. Of greater significance are extensive areas of flood basalt flows, usually called large igneous provinces (LIPs), that occur on land and in the oceans. Because basalt lava is quite runny (i.e. has low viscosity) it is able to flow for many hundreds of kilometres forming flat, sheet-like flows rather than the conical edifices of volcanoes (Figure 13). The stacked layers of lava flows usually weather to form a stepped topography often called traps.

The environmental impact of LIPs is due to their huge size and the speed of their eruption. Lava volumes range from half a million to several million cubic kilometres, and individual flows may have volumes that approach 10,000 cubic kilometres. This is a scale of volcanism unparalleled today; the biggest eruptions of historical times have never emitted more than 20 cubic kilometres. It is fortunate that there are no active LIPs today. The last one, called the Columbia River Traps, formed around fourteen million years ago in the north-western United States and was a very small example being less than 0.2 million cubic kilometres in size. As well as their scale, it is the short duration of eruption that is also likely a significant factor in LIP-triggered environmental damage. Many of them seem to have erupted the majority of their lavas in <1 million years. The coincidence of LIP eruptions with nearly all extinction events is a further compelling reason to link these two phenomena (Table 1).

The actual connection between LIP volcanism and mass extinction is not immediately clear. The main effect of modern volcanic eruptions on climate is primarily short-term cooling because of the injection of dust and gases into the atmosphere (Figure 14). Sulphur dioxide (SO_2) is one of the main volcanic gases and this reacts with water vapour to form clouds of sulphate aerosols which reduce the amount of sunlight reaching the Earth's surface. However, this effect is very short-lived because the aerosols (and dust) are rapidly removed by rainfall and so the cooling effect, often dubbed a 'volcanic winter', only persists for a year or so after the eruption has ceased. It is

Table 1. Comparison of the proposed extinction mechanisms for all mass extinctions

Cause of extinction	LIP	Meteorite	Cooling	Warming	Ocean anoxia	Ocean acidification	Ozone depletion
Early Cambrian	√				√		
End Ordovician	√?		√	√	√		
Late Devonian	√?	√?	√		√		
End Devonian			√		√		
Mid-Permian	√		√?	√	√?		
P-Tr	√			√	√	√	√
End Triassic	√			√	√?	√	
Early Jurassic	√			√	√	√?	
End Cretaceous	√	√	√			√?	

probable that some of the largest basalt flows of LIPs may have erupted over several decades, thus prolonging the cooling duration, but even this may not have been enough to trigger long-term climate change.

Modelling the effects of giant eruptions has suggested that global temperatures could briefly drop by up to 8°C during the lifetime of the eruption but are unlikely to have been more severe than that (Figure 14). This is potentially significant for life on land but unlikely to be catastrophic, and the thermal heat capacity of the oceans is so great that such brief cooling is unlikely to have any effect on this realm. There is also an interesting non-linear response of the atmosphere to the injection of SO_2. As more SO_2 is erupted, the size of the sulphate aerosol droplets increases, which acts to decrease the effect of sunlight scattering. So, while

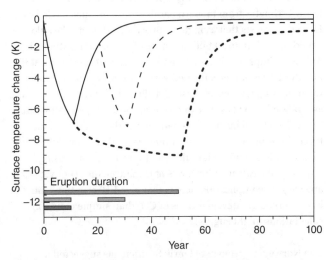

14. **Computer model of the global cooling effects of sulphate aerosols produced by large basalt eruptions of the Deccan Traps. Three scenarios are modelled, a single 500-cubic-kilometre flow, two separate 500-cubic-kilometre flows separated by ten years, and a single 1,000-cubic-kilometre flow erupted over fifty years.**

larger eruptions produce more aerosol particles, which increases cooling, the larger size of droplets weakens the effect. In other words, as volcanic eruptions become larger the cooling effect becomes self-limiting, making it unlikely that they could ever trigger an environmental catastrophe.

After a few years, the short-term cooling effect of an eruption is replaced by a longer term warming trend because of the release of CO_2, a greenhouse gas. As we know today, CO_2 emissions are not immediately removed from the atmosphere; they build up over time and are only slowly removed by processes such as weathering of rocks and burial of organic matter and carbonate in sediments. This means that if LIP eruptions are reasonably closely spaced, say every few hundred years, then the warming effect of the CO_2 releases becomes additive because there is insufficient time for removal. So, how much gas is released? Using estimates from modern basalt eruptions, such as those on Hawaii, suggests that a single 1,000-cubic-kilometre lava flow would release up to 13 billion tonnes of CO_2 into the atmosphere. This is an impressive amount, and the observation that most if not all LIP eruptions coincided with episodes of rapid global warming suggest a clear cause and effect. However, it is not quite as straightforward as that when we consider CO_2 emission rates today. Fossil fuel burning annually currently releases around three times more CO_2 into the atmosphere than a single 1,000-cubic-kilometre lava flow, and yet this is not causing anything like catastrophic levels of climate change (not yet, anyway). The implication here is that past LIP eruptions must have erupted greater quantities of CO_2 than simple emissions calculations might suggest.

The Norwegian geologist Henrik Svensen has suggested a possible solution to the conundrum of how to make volcanism more climatically significant. Studying several LIPs, he has identified features called 'breccia pipes' in the surrounding areas.

These are vertical zones of fragmented rock that he attributes to the violent escape of gas derived from heating sedimentary rocks. As magma rises to the surface it heats the surrounding rocks in a process called contact metamorphism. If the rocks contain substantial amounts of organic carbon, or if they are limestones (calcium carbonate), then this 'baking' process produces large amounts of CO_2. Svensen's hypothesis therefore suggests that the rapid generation of thermogenic gas may greatly increase the total CO_2 emissions from LIP volcanism and so drive extreme warming. The steps that are then required to get from global warming to mass extinction are outlined in the next section and summarized in Figure 15.

CO_2 and SO_2 are not the only gases released by volcanism; halogen emissions (chlorine and fluorine) are also substantial. These gases produce acid rain, although modelling suggests any environmental damage from this would be minor. More importantly, halogens hinder the formation of ozone leading to the possibility that Earth's stratospheric ozone shield, which protects life from harmful UV-B radiation, may be damaged during LIP eruptions. Other gas emissions may also contribute to ozone destruction. Some LIPs, such as the Siberian Traps, erupted through substantial coal and salt deposits which, when baked, would release large amounts of chloromethane (CH_3Cl) in addition to the CO_2 noted above. If eruptions are violent enough to inject this potent gas into the stratosphere, then it could cause severe ozone destruction.

Large-scale volcanism is clearly capable of a multiplicity of environmental and climatic effects because of the range of gases emitted and their diverse consequences. Mass extinction mechanisms involving LIPs thus emphasize a cascade of cause and effects often summarized in a flow chart (Figure 15). The nature of these effects and the evidence that they were actually manifest in the past is described in the following sections.

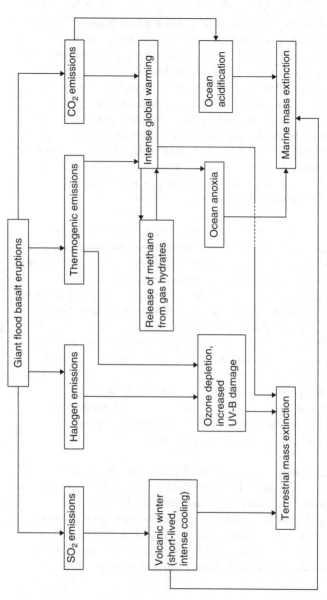

15. **Flow chart showing how large volcanic eruptions can ultimately lead to mass extinction via several steps.**

Hyperthermals

Earth's climate is constantly changing, either getting warmer or cooler, but several of the major mass extinctions, notably the P-Tr and end-Triassic events, and the early Toarcian crisis, coincided with especially rapid rates of warming. We know this from our ability to assess past temperature using the proportions of oxygen isotopes preserved in fossil material. This has been done using shells made of calcite ($CaCO_3$), but bone (calcium phosphate) is generally the preferred material nowadays. The oxygen isotope ratio of bone and calcite is set during growth and is mostly controlled by the temperature at the time of formation. As long as the material does not change its chemistry during burial (bone is thought to be especially resilient to such changes) then past temperatures can be calculated.

Studies of phosphate oxygen isotope ratios during the P-Tr mass extinction indicate that sea surface temperatures rose by an impressive 12°C in a few hundred thousand years and may have reached 35°C in equatorial latitudes. For comparison, the warmest oceans today rarely exceed an annual average of 28°C. Year-round temperatures in the mid-30s °C is inimical to the formation of much marine life, and the selectivity of P-Tr extinction losses may have been caused by these super-hot ocean temperatures. The modern-day planktonic radiolarians are intolerant of temperatures above 30°C, and the near elimination of this group at the end of the Permian is strongly suspected to have been caused by the high temperatures.

The distribution of the survivors in the aftermath of the P-Tr extinction also seems to reflect the high temperatures, especially in equatorial latitudes. Amphibians are a quintessentially warmth-loving equatorial group, throughout nearly all their considerable 360-million-year history. In dramatic contrast, amphibians thrived during the earliest Triassic, but only at high latitudes, and they were absent from lower latitudes where temperatures were presumably lethally hot.

There has been considerable debate as to how the world could get so hot so quickly during these mass extinctions. As described above, volcanic emissions, exacerbated by thermogenic sources of CO_2, no doubt played a role but these may have triggered the release of further greenhouse gases. In the deeper, colder parts of the oceans today large amounts of methane are found beneath the seafloor where it occurs bound in the crystal lattice of ice. These deposits are called gas hydrates or methane clathrates, and they exist at a delicate threshold where an increase of bottom water temperatures of only a few degrees would be enough to release the methane as the ice melted. Methane is a highly effective greenhouse gas (although it rapidly oxidizes to CO_2 in the atmosphere) and so the destruction of hydrates could lead to a much more intense greenhouse climate. There is indirect evidence that this feedback may have happened during several mass extinctions, but this requires a knowledge of the carbon isotope ratios in sediments.

Carbon atoms occur as two isotopes—carbon-13 and carbon-12 (there is also carbon-14, but this is radioactive and does not survive in sediments for long, so it does not concern us here). Most organic matter carbon is relatively enriched in carbon-12, compared to the inorganic forms of carbon dissolved in seawater, while the carbon found in limestone tends to have an isotopic ratio like the seawater from which it formed. The isotope ratio is expressed as the concentration of carbon-13 over carbon-12 ($^{13}C/^{12}C$). Organic carbon generally has a 'light' ratio compared with limestone because it has more ^{12}C making the ratio a smaller number than that for carbon-13-rich limestone.

Now, during many mass extinctions, including the P-Tr example, the $^{13}C/^{12}C$ ratio of both organic carbon and limestone becomes lighter. If it was just the organic matter that changed its ratio at these times then a whole range of processes could be involved. It might be that there had been a change in the plankton which have different abilities to select the light carbon-12 for their organic matter. Fortunately, because these changes affect both the organic matter and the limestone, the cause is more likely simpler than that.

Many geologists have explained the trend to light carbon isotope ratios as a signature of methane release from gas hydrates into the ocean and atmosphere because this gas has a very light $^{13}C/^{12}C$ ratio. Once in the atmosphere, methane acts as a greenhouse gas, and over a period of a few years it oxidizes to CO_2. Exchange of this gas between the atmosphere and ocean waters ensures that some is used during photosynthesis by plankton and some is used by the organisms that form limestone. This leaves the tell-tale carbon-12-rich signature in geological strata.

If methane release from hydrates was the only mechanism that altered the ocean's $^{13}C/^{12}C$ ratio then this would be clear evidence for a methane-driven greenhouse during mass extinctions, but unfortunately this is not the case, because there are other ways of making the ratio lighter. Volcanic CO_2 also has a light $^{13}C/^{12}C$ ratio (although not to the same extent as methane), and so major eruptions will affect the value. Changes in the abundance of organic matter in the world also affects $^{13}C/^{12}C$ ratios and so there is plenty of scope for debate on what causes carbon isotope changes. Furthermore, none of the mechanisms are mutually exclusive, and several factors were probably operating during mass extinctions making it difficult to distinguish their relative influence.

Ocean anoxia

Development of extensive ocean anoxia is the most common environmental change encountered during mass extinction events (Table 1), with the K-Pg crisis being one of the few exceptions. To put the significance of this observation into context, it is important to stress that anoxic conditions are very rare in today's oceans and are only encountered in a very few, limited areas. This has also been the case for most of the past 600 million years, the oceans have been well-ventilated with oxygen available at all water depths. This is because ocean circulation is generally vigorous (it is driven by density differences caused by salinity and temperature variations) and able to transport well-oxygenated surface waters into the deepest ocean settings. To a large extent, the circulation is controlled by the equator-to-pole temperature gradient and it may be that this weakens during global warming because polar regions warm up relatively more than equatorial ones.

There are several other links between warming and ocean anoxia. Higher temperatures favour low oxygenation because less oxygen dissolves in warmer waters. Also, the decay of organic matter produced by plankton speeds up as temperature rises, which consumes more dissolved oxygen in the water column. This is why we keep food (i.e. organic matter) in a refrigerator in order to slow down its decay. Finally, the amount of organic matter in the oceans also controls oxygen levels because its decay uses oxygen. This is controlled by the amount of nutrients available to plankton which is likely to increase as the world gets warmer. This is because warmer climates are associated with more rainfall which leads to increased supply of nutrients such as phosphates and nitrates via run-off from the continents. More nutrients mean more organic matter production (plankton blooms) which in turn leads to more decay and oxygen consumption. All these factors are likely to act in combination when the oceans become anoxic.

Anoxic oceans cause major problems for marine life by making large areas uninhabitable because oxygen is required for respiration and because of the poisonous effects of dissolved gases such as hydrogen sulphide which build up in the absence of oxygen. Marine anoxia appears to have been especially intense during the end-Ordovician, Late Devonian, end-Devonian, P-Tr, and the Early Jurassic extinctions (Table 1). However, the link between anoxia and extinctions is not perfect: there have been other intervals of ocean anoxia which are not associated with mass extinctions. One of the best known occurred around ninety million years ago in the Late Cretaceous. This has been called the Bonarelli event and, although it is associated with a few losses, these are distinctly modest compared with true extinction episodes.

The intensity of extinction may relate to the extent of anoxia within the water column. Bonarelli anoxia only appears to have developed in deeper parts of the water column below a few hundred metres depth and so plenty of well-ventilated, shallow sea habitats remained at this time. In contrast, anoxia during other mass extinctions, such as the Late Devonian and P-Tr events, developed through most of the water column, from the deep ocean floor into exceptionally shallow waters, leaving little room for marine life. It has even been proposed that such episodes of intense marine anoxia may have caused extinctions on land because hydrogen sulphide, built up in the oceans, may have leaked into the air and drifted inland. However, this toxic gas oxidizes very quickly in the atmosphere (to SO_2) and so any potential effects are unlikely to have extended much beyond coastal environments.

Ocean acidification

The modern oceans have an alkaline pH and it is likely that this has been so since complex life first evolved and started secreting carbonate skeletons. However, the rapid introduction of large amounts of CO_2 into the atmosphere causes ocean surface waters to acidify, as the CO_2 is dissolved, potentially making it more difficult to secrete carbonate. This is a concern in modern oceans where ocean acidification has been mooted as an impending crisis for many shellfish, especially in higher latitudes because colder waters can dissolve more CO_2. It has also been suggested as a major factor in several mass extinctions, especially the end-Triassic event. Evidence for acidification-driven extinction at this time includes the preferential loss of aragonite-shelled molluscs (aragonite is more soluble than the alternative calcite) and a general rarity of limestone deposits during the extinction. Claims for acidification stresses during other mass extinctions have also generally focused on preferential losses among thick-shelled species. However, other factors could also be at play here: thick-shelled forms tend to dominate in shallow waters, which are more susceptible to higher temperatures, for example. There have been attempts to develop geochemical proxies for ocean pH although none are, as yet, very reliable. Ocean acidification is currently difficult to test for as an extinction mechanism.

Ozone depletion

Destruction of the ozone shield followed by intense ultraviolet (UV) irradiation is a potent extinction mechanism, particularly for terrestrial plant life. It has been suggested as the cause of the devastating floral extinctions during the P-Tr mass extinction. However, proving such a scenario is currently difficult because there is no fossil proxy for UV radiation available although several are being developed. Indirect clues, such as the mutant pollen seen during the P-Tr crisis, may indicate UV damage although this evidence really only points to the fact that plants were suffering some form of stress at the time.

Ice ages

Despite present concerns about global warming, the Earth is in an ice age characterized by phases of dramatic ice advance and warmer, interglacial interludes. These conditions began around 2.5 million years ago and, despite dramatic, large-scale environmental changes, there has been no associated extinction crisis (except for selective extinctions among large animals that began around 40,000 years ago; this is discussed later). Despite the apparently benign nature of our current ice age, cooling and glaciation have long featured in extinction scenarios, especially for the end-Ordovician and end-Devonian mass extinctions. Interestingly, these are the two mass extinctions where the losses were especially intense among pelagic groups, suggesting that changes in ocean circulation and water column structure driven by cooling may have been important. There is ample evidence for cooling during both these intervals in the form of glacial deposits found in polar latitudes. The selectivity of the first phase of the end-Ordovician extinction, with tropical marine life being especially hard hit, is also suggestive of a cooling-driven crisis. The Mid-Carboniferous decline in diversity also seems strongly tied to global cooling and glaciation. What remains to be answered is why these ancient glaciations were so much more lethal than the most recent ice age. If major oceanographic changes drove the past mass extinctions, then maybe the current ocean circulation regime is more robust and resilient to temperature-driven perturbations?

Sea-level change

Like global temperature, global sea level (eustasy) is always changing. However, during most mass extinctions sea-level changes are often rapid and of high amplitude. Thus, a lot of early studies, from the 1950s to 1990s, considered them responsible for the crises. For example, it was suggested that major sea-level fall in the Late Permian saw the area of shelf seas shrink as the coastline retreated to the continental margin, causing the loss of shallow marine habitat space, and thereby leading to extinctions. Clearly, this extinction-by-habitat-loss only applies to marine life because land area increases during regression. More recent study has overturned the link between regression and the P-Tr mass extinction, and shown that the losses occurred during a phase of sea-level rise *after* a low point of sea level. In fact, many mass extinctions occurred during transgression at times of global warming when marine anoxia spread across shelf seas. It is these two factors, warming and anoxia, rather than the associated sea-level change, that are usually implicated in current extinction models. Few geologists now consider sea-level change to be directly implicated as a cause of mass extinction.

Meteorite (and comet) impact

The idea that mass extinction could be caused by the impact of meteorites (also known as asteroids or bolides) or comets dates back to the mid-20th century but the lack of proffered evidence meant that few believed the claims. It was only with the publication of a paper in 1980 by a group led by the father and son team of Luis and Walter Alvarez that this ultra-catastrophist idea gained rapid acceptance. Ironically, Alvarez and company used one of the most obscure ways of detecting impact—they measured the concentration of iridium in boundary sediments at the level of the Cretaceous-Paleogene mass extinction. This metal is very rare in the Earth's crust but relatively abundant in meteorites. The Alvarez team showed iridium levels reaching concentrations of several parts per billion at the boundary, compared to normal background levels of <30 parts per trillion: an anomaly best explained as a sudden influx of extra-terrestrial material.

Following the iridium discovery, additional evidence began to accumulate quickly. Shocked quartz is the name given to grains of this mineral showing cross-hatched deformation bands (normal quartz is clear) that are produced at ultra-high pressures during meteorite impact, or at nuclear test sites today. These grains, together with stishovite, a high-pressure variant of quartz, were found in K-Pg boundary sediment. In addition, microtektites and microspherules turned up. The former represent rapidly cooled droplets of liquid rock and the latter are condensed blobs of vaporized rock, and are both examples of material blown out of an impact crater.

Ten years after the initial discovery of the iridium anomaly, the most irrefutable evidence of all, a giant impact crater

180 kilometres in diameter and sixty-six million years old, was located buried beneath younger rocks at Chicxulub on the Yucatán Peninsula in south-east Mexico. This is the largest known crater on Earth, and the regional geology reveals that the target rocks were mostly limestone and anhydrite (a mineral composed of calcium sulphate formed by evaporating seawater), although some sandstone must also have been present to source the shocked quartz.

The K-Pg extinction scenario envisages a 10-kilometre-diameter meteorite hitting the Earth, releasing energy equivalent to 100 million megatons of TNT (this is around a billion times more energy than was released by the atomic bomb dropped on Hiroshima) and blasting huge amounts of material (ejecta) into the atmosphere. Initial ideas focused on the cooling effects of dense clouds of dust, but modelling and comparison with large-scale volcanic eruptions indicates that this phenomenon would be very short-lived (a few months) because dust falls out of the atmosphere quickly. Cooling could also have been exacerbated by soot from burning triggered by hot ejecta material raining back to the ground and setting fire to vegetation. However, probably the most harmful effects came from the vaporization of the Yucatán target rocks. Anhydrite would produce SO_2, and ultimately cooling clouds of sulphate aerosols, while the limestones would produce CO_2. As it happens, these are the same two gases implicated in the volcanism-driven extinction models described above. So, the Chicxulub impact can be thought of as a super-sized eruption, with intense darkness and cooling caused by sulphate aerosols (aided by soot?) lasting for several years, followed by warming due to the greenhouse gas effects of the CO_2. The abrupt extinction of many photosynthesizing organisms at the K-Pg boundary, especially calcareous plankton, suggests that a period of darkness, and shutdown of photosynthesis, is likely to have been the single most devastating consequence of the impact.

Despite the amassed evidence, there have been frequent challenges to the Alvarez hypothesis. Some have claimed that the Chicxulub crater is too old to be implicated in the K-Pg crisis, even though the iridium anomaly occurs precisely at the level of mass extinction. The role of the Deccan Traps flood basalt eruptions in India, which reached their peak intensity exactly at the boundary, also provides an additional complication and source for debate. Nonetheless, the K-Pg impact scenario has been well-established in the scientific community since the 1980s. Much more controversial is the idea that other mass extinctions, and maybe even all extinctions, can be attributed to impact. David Raup, a Chicago palaeontologist, was the most ardent proponent of this viewpoint which he put forward in a 1991 book *Extinction: Bad Luck or Bad Genes?* The title asks whether species go extinct because they are poorly adapted (essentially a Darwinian view), or because of bad luck, like a meteorite strike for which no amount of inbuilt genetic resilience can prepare species. He supported his arguments with a 'kill curve' that correlates extinction intensity against scale of impact (Figure 16). This conceptual graph suggests that very rare giant impacts, such as the Chicxulub example, are likely to cause mass extinctions on the scale of the K-Pg catastrophe, while more frequent smaller impacts also cause extinction but on a smaller scale. If the kill curve works, then it should be possible to show that other mass extinctions coincided with meteorite (or comet) strikes.

Ever since 1980, there has been a prolonged search for impact evidence, such as iridium anomalies and shocked quartz, at other mass extinction horizons, but after forty years the amassed results are meagre. Conversely, there has been a much more successful search for craters, but, unfortunately for impact-extinction proponents, they do not coincide with any major crises. The Manicouagan Crater in Ontario is, after the Chicxulub crater, one of the largest known examples. It is 100 kilometres in diameter and dates to a level in the middle of the Late Triassic, 216 million

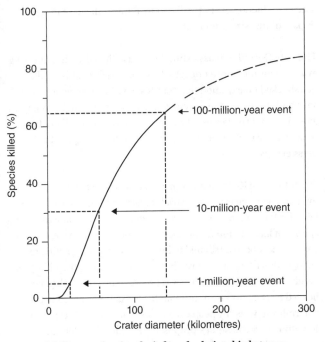

16. **Raup's kill curve showing the inferred relationship between the number of species killed and the size of the impact crater. Large-impact events are much rarer than small ones. Note that the 180-kilometre-diameter crater at Chicxulub is credited with ~70 per cent species extinction, the only known impact–extinction link.**

years ago. There was no associated mass extinction, nor even a minor crisis, at this time. A slightly better impact–extinction link comes from the late Eocene Epoch thirty-six million years ago when iridium anomalies, microtektites, and shocked quartz all turn up in the sedimentary record. There is even an impact crater of the right age, the 100-kilometre-diameter Popigai structure in Siberia. Some extinction losses, especially among calcareous plankton, occurred in the late Eocene and so could be potential victims, but these seem to be spread over several million years. There is certainly no evidence for a >40 per cent species extinction event predicted by Raup's kill curve for an impact of this size (Figure 16).

The Late Devonian mass extinction is also linked with tantalizing evidence for meteorite impacts. Iridium enrichment is weak, but shocked quartz and microtektites have been found at a few sites and there are at least four impact craters of roughly the right age. However, the largest of these, the 55-kilometre-diameter Siljan Crater in Sweden, is rather small to have triggered a mass extinction.

Other than the K-Pg event, perhaps the most spectacular claim for a link between extra-terrestrial impact and extinctions is for a late Pleistocene event (12,800 years ago), around the time that many species of large mammals disappeared. This megafaunal extinction has generally been attributed to human hunting and/or climate change (described in the next chapter), but in 2007 an exciting and highly contentious alternative was proposed in a research paper headed by Richard Firestone (a fantastic example of nominative determinism whereby people have jobs that relate to their name). Firestone and colleagues proposed the following idea: the megafaunal extinction happened when a giant comet disintegrated and exploded in the upper atmosphere blasting the land below with intense heat that ignited vast tracts of vegetation, especially in the northern hemisphere. The resultant clouds of soot caused intense cooling and triggered a short-lived mini ice age called the Younger Dryas event. Placing this climatic event in context, the last peak of northern hemisphere glaciation occurred around 22,500 years ago and ice sheets began a rapid retreat 20,000 years ago. The Younger Dryas cooling was a strange, intense cooling episode, lasting around 1,800 years, within the current interglacial warm period. The rapid onset of the Younger Dryas cooling suggests a comet-driven cooling hypothesis is at least plausible.

Potentially, the extinctions of many large ice age mammals may have been caused by an episode of instantaneous cooling exacerbated by the loss of vegetation due to wildfires. The human population of North America may not have fared much better. Contemporary populations belonged to the Clovis Culture, named after distinctive spear points found across the North American continent. These disappear abruptly from the archaeological record at the onset of the Younger Dryas. The reappearance of archaeological artefacts after the Younger Dryas show much more regional differentiation into separate cultures and so point to a major change in human settlement patterns.

Firestone's original (2007) evidence for detonation of a giant comet in the atmosphere above North America included a widespread black layer, suggested to represent the soot from the burnt forests, in addition to iridium-bearing grains and microspherules. There was initial controversy when follow-up investigations failed to verify some of these findings, but the evidence has continued to accumulate. Thus, ice cores from Greenland show enrichment of ammonia, a product of burning, at the onset of the Younger Dryas, while contemporaneous enrichments of platinum and osmium suggest a sudden input from cosmic sources like iridium. The evidence for a giant cometary explosion is therefore compelling (although still challenged by many), and the discovery in late 2018 of a crater (31 kilometres in diameter) in north-west Greenland may provide an impact site of a comet or meteorite of the right age. Whether this event drove extinctions is less clear though, because the last occurrences of the large mammals are not so directly tied to the event. North American megafaunal extinctions straddle the Younger Dryas period with some distinctive animals such as the woolly mammoth (*Mammuth primigenius*) surviving until 10,000 years ago, a considerable time after the Younger Dryas cold snap. Only a handful of species seem to have gone extinct at the precise time of impact. Spectacular it may have been, but the cometary impact/explosion does not appear to have played a major role in the Pleistocene extinction losses. These are discussed further in Chapter 6.

Chapter 6
What happened to the Ice Age megafauna?

The Ice Age or Pleistocene Period, from 2.6 million to 11,650 years ago, was a time when the climate cycled from glacial to interglacial states every 100,000 years or so. The consequent waxing and waning of continental ice caps caused dramatic changes in shelf sea area as global sea level fluctuated by up to 200 metres. Many islands became regularly connected and then isolated from other land masses, a process that caused the repeated isolation and then mixing of endemic faunas. The immense diversity of the island-rich Southeast Asian region has been attributed to this process, which is likely to have fostered the evolution of new species (or speciation). However, despite the magnitude and speed of all these climatic changes that required frequent, rapid migrations by most species, the Pleistocene was not a time of unusually high extinction rates in either marine or terrestrial settings. The sole exception occurred during and after the last glacial maximum when numerous large, terrestrial animals disappeared. This has been termed the *Pleistocene megafauna extinction* and its cause has been one of the most highly contested topics in palaeontology for the past fifty years.

Often called a 'mass extinction', this term is not really appropriate to the megafauna extinction because losses overall were trivial compared to the scale experienced during the big five mass extinctions. They were also highly selective. True mass extinctions by definition are global-scale events with extinctions spread among diverse ecological groups and habitats. The megafauna are usually defined as animals with greater than 44-kilogram mass (although some use a 100-kilogram definition) and, using this somewhat arbitrary definition, it appears that ninety-seven genera (nearly two-thirds of the global total), and nearly 180 species disappeared late in the Pleistocene. In fact, analysis of the losses has shown that the extinctions preferentially removed slowly reproducing species, especially those that produced fewer than one offspring per female per year. Such animals tend to be large species hence the 'megafaunal' nature of the crisis. However, some small species also reproduce slowly and these too suffered high extinction levels, such as several extinct species of spiny anteaters in Australia which only weighed around 7 kilograms. Those large animals that survived the cull tended to be in Africa where a highly diverse megafauna still roams as it has since the Pleistocene; or in the dense forests of Southeast Asia which has retained elephants, rhinos, and large apes—although all are now endangered by modern habitat loss and hunting.

The extinctions occurred from 40,000 to 3,000 years ago, but the intensity and timings varied from continent to continent (Figure 17) and, as a result, separate discussions have developed for each region with different alternatives holding swaying. Nonetheless, the selective nature of the losses is very similar on all continents, and the same two culprits are always in the dock: climate change and humans.

At the start of the extinctions, the world was in a relatively warm, interglacial period, but the climate gradually cooled into glacial conditions that peaked 22,000 years ago. This was followed by temperature rise that culminated in a warm interval 14,700–12,800 years ago known as the Bølling-Allerød Interstadial (Figure 17). The warmth was briefly interrupted by the Younger Dryas cool episode (12,800 to 11,600 years ago), but it rapidly returned (this point marks the Pleistocene/Holocene boundary) into the

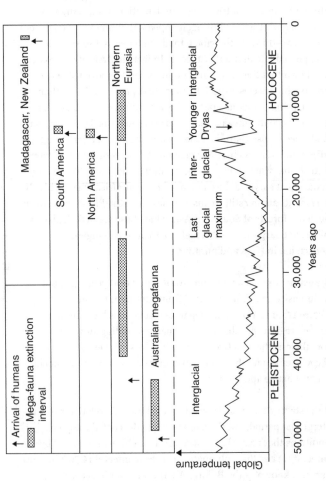

17. Timing of Pleistocene megafaunal extinctions showing the variation from continent to continent compared with temperature change and the arrival of humans (arrowed).

interstadial which we are still enjoying today. With the possible exception of the Younger Dryas, none of these climate changes were in any way unusual, there had been dozens of similar oscillations throughout the Pleistocene.

The unique aspect of the late Pleistocene was the migration of modern humans out of Africa and their spread across the globe. As we shall see, our species' arrival on different continents was often closely followed by megafauna extinctions (Figure 17). This correlation suggests a causal link, and it was first proposed fifty years ago by Paul Martin, an anthropologist from the University of Arizona. He suggested the extinctions were human-driven from intense hunting, and termed the arrival of humans a 'Blitzkrieg' (a military term for 'a lightning war'). Humans are also capable of causing intense environmental changes such as deforestation which may inadvertently lead to extinctions (this sometimes gets called 'Sitzkrieg'). The megafauna of the African continent escaped all this war-like strife and they provide a useful counterpoint to *Blitzkrieg/Sitzkrieg* models. Martin argued that African mammals spent hundreds of thousands of years co-evolving with human species and so learned to be wary of them. In contrast, the megafauna of other continents likely had a naïve appreciation of the dangers they faced from the newly arrived, innocuous-looking humans.

Right from the beginning, Martin's hypothesis was controversial with many scientists downplaying the role of humans in megafaunal extinctions and instead pointing to climate change and its role in shifting habitats. There were certainly many fluctuations during the extinction interval, including the rapid onset of the Younger Dryas cooling. The Pleistocene climate was also punctuated by numerous Dansgaard–Oeschger Events. These were brief (~1,000 year) warm episodes superimposed on cooler intervals, and have been proposed as the main cause of extinctions. However, the correlation precision is not yet quite good enough to test this link, and it is important to remember the extinctions occurred at widely varying times on different continents; as such, no single climatic event could be responsible for the extinctions.

Australia

Much of the debate on the megafauna extinctions is caused by the variable quality and quantity of dating evidence for human arrival and extinctions. This is especially the case for the losses in Australia which experienced the first and most intense of all the late Pleistocene crises. Extinctions appear to have begun rapidly around 45,000 years ago when twenty-three out of the twenty-four terrestrial animal genera that weighed more than 45 kilograms disappeared in fewer than 2,000 years (Figure 18). The main victims were large marsupials including *Phascolonus*, a 200-kilogram wombat; *Diprotodon*, a giant herbivore weighing more than 2 tonnes; *Procoptodon*, a giant kangaroo; *Protemnodon*, a giant wallaby; and *Thylacoleo*, a 130-kilogram marsupial lion. Reptiles were the top predators in the Australian landscape and these too went extinct. They included an impressively large lizard, *Megalonia*, which was a close relative of the Komodo dragon albeit bigger at more than 5 metres in length and weighing in at >600 kilograms. In more forested areas, a gigantic snake, the 9-metre-long *Wonambi*, may have taken the top predator role. The 90-kilogram red kangaroo was the sole survivor of this diverse assemblage of fascinating creatures. Tasmania was connected to mainland Australia at this time and this region appears to have provided a temporary refuge for some animals. Thus, some of the marsupial megafauna, such as *Thylacoleo* and *Protemnodon*, survived until 41,000 years ago in this southernmost outpost.

Human arrival time in Australia is not precisely known, but was probably sometime around 50,000 years ago, shortly before the extinctions began; although there is tentative evidence for a much earlier arrival around 65,000 years ago. They arrived by boat from the north and spread their way south through the continent. Thus,

18. Examples of extinct Pleistocene megafauna of Australia. Clockwise from top left *Diprotodon*, *Megalania*, *Thylacoleo*, and *Phascolonus*.

the spread in timing of the extinctions, with the last survivors occurring in Tasmania, is strongly supportive of an overkill model. Nonetheless, climate change also has its supporters with gradual cooling and aridification blamed for the losses. Part of the problem in resolving debates on the extinction of the Pleistocene megafauna stems from the general rarity of terrestrial tetrapod fossils and thus the problem of precisely dating first and last appearances—especially of a species like *Homo sapiens*, which has a very scrappy fossil record. A partial solution to this problem comes from the high-quality, prolifically abundant spore and pollen fossils found in sedimentary rocks, which provide a high-resolution history of plant communities. The key fossil is a spore (*Sporormiella*) produced by a fungus that decomposes the dung of herbivorous animals. In a landscape covered in the dung produced by abundant plant eaters, *Sporormiella* is very common. This was the case in Australia up until 45,000 years ago, but the spore became much rarer over the subsequent 2,000 years indicating the rapid disappearance of most herbivores and their dung. This decrease does not coincide with any climate change, but rather it occurs shortly after human arrival. Human *Blitzkrieg* is thus a strong contender for the elimination of the Australian megafauna, unless the 65,000-year arrival date can be confirmed.

Northern Eurasia

The disappearance of the Northern Eurasian megafauna is the most complex of all the late Pleistocene crises because the losses were staggered and consisted of both regional extinctions (or extirpations) and true extinctions. They can be broadly grouped into two main phases that occurred before and after the last glacial maximum, with the earlier one spread over a considerable period of time, from 40,000 to 22,000 years ago. Our own close relative, *Homo neanderthalensis*, was among the first to go during this phase, while the straight-tusked elephant *Palaeoloxodon* and the cave bear *Ursus spelaea* were the last. Strictly speaking, *Palaeoloxodon* was only extirpated in mainland Europe because a population may have survived in China until 3,000 years ago, and several dwarf species lived on Mediterranean islands until the same cut-off time.

Extinction losses intensified during the second phase and were concentrated in a shorter interval, from 13,000 to 10,000 years ago, that encompasses the rapid climatic oscillation of the Younger Dryas. However, it is difficult to tie any species losses precisely to the sudden onset of this cold episode. The closest correlation is the extinction of the woolly rhinoceros (*Coelodonta antiquitatis*) the fossils of which became increasingly rare after 15,000 years and is last seen depicted in cave art 12,500 years ago. It was outlasted by the iconic woolly mammoth, one of the last megafauna species to go extinct in Eurasia. Their disappearance shows interesting regional variations and staggering across time periods. In Europe they became very rare after 15,000 years, but, like the woolly rhinoceros, they are still seen depicted in human artefacts (drawings etched in bone) from Germany until 12,000 years ago. It appears that their range gradually contracted northwards, because the youngest mainland mammoths survived in the Taimyr Peninsula of northern Siberia until 9,650 years ago. A small population of mammoths even held out on Wrangel Island in the Arctic Ocean where the last individuals died only 3,000 years ago, meaning that they survived into the late Bronze Age.

Total megafaunal extinctions in Eurasia were much fewer than in Australia, but they still represent a significant crisis. Linking the losses to climate change is not easy because they occurred over a broad interval that encompassed the full spectrum of Pleistocene climates, and the ecology of the animals varies greatly. Thus, the grass-feeding mammoth lived in tundra habitats, while the straight-tusked elephant was a browser that lived in broad-leaved forests. Extinction scenarios thus have to be individually tailored for each species, making for untidy models. Climate-driven extinction scenarios point to factors like habitat fragmentation caused by either cooling or warming trends; but of course none of these changes was unusual in the context of earlier Pleistocene climatic variations. Furthermore, extensive areas of tundra and forests, such as the taiga of Siberia, remain to this day.

Did the migration of modern humans out of Africa cause the Eurasian megafaunal extinctions? They first turned up in central Eurasia 50,000 years ago and reached western Europe around 5,000 years later. However, with the exception of the Neanderthals' extinction (discussed below), these arrival dates considerably predate the megafaunal extinctions (Figure 17). It has been argued that the Eurasian megafauna had lived alongside earlier human species, such as the Neanderthals, for hundreds of thousands of years, and so were to some extent better adapted to cope with human hunting pressures, possibly because of the former's relatively high reproductive rates. Eventually technological advances, such as the invention of bows and arrows and spears around 18,000 years ago, may have intensified hunting pressure and so caused the second, most intense phase of megafaunal extinctions over the next few thousand years. Further support for an overkill hypothesis comes from the gradual retreat of the mammoth north-eastwards, closely mirroring the expansion of human populations. Archaeological evidence suggests people finally reached Wrangel Island 2,700 years ago which roughly corresponds with the demise of the last mammoths on this Arctic refuge.

The Americas

The human colonization of the Americas began from the north around 14,500 years ago, with migrants heading southwards through Canada as continental ice sheets retreated, opening a corridor to the east of the Rockies. By 13,500 years ago the Clovis people and their eponymous stone spear points were found throughout the United States. There have been controversial claims for pre-Clovis archaeological sites up to 15,600 years old, especially in coastal locations, and it is possible that some people arrived earlier by boat. However, the Clovis people were clearly the first to establish a continent-wide dominance across North America—albeit for only a short time, because their spear points disappeared at the start of the Younger Dryas 12,800 years ago. Clovis hegemony closely coincides with a catastrophic extinction of the North American megafauna: thirty-five mammal genera disappeared in around 1,500 years, including the mastodon (*Mammut americanum*), the woolly mammoth, and giant beavers (*Castoroides*) that were 2 metres long and up to 125 kilograms in weight, while camels and horses were extirpated from the continent. Like the Australian megafaunal extinction, the

North American losses coincided with a crash in the abundance of *Sporormiella*, marking the eradication of megafaunal herbivores and their associated dung from the landscape.

There was a similarly intense megafaunal extinction in South America that removed animals such as *Megatherium*, a giant ground sloth that weighed 4 tonnes; *Glyptodon*, a giant armadillo; and a native horse (*Hippidion*). The South American losses appear to have occurred later (around 11,500–8,000 years ago) than those in North America, and they began with the arrival of the first humans. Clearly, the temporal link between extinctions and human arrivals in the Americas is very close. This is exemplified further by the numerous losses among the sloths, a family characterized by low reproductive rates and a range of body sizes that peaked with *Megatherium* but also includes tree-dwelling, two-toed species that weigh only 6 kilograms. North American sloths disappeared over 13,000 years ago, South American sloth losses were nearly 2,000 years later, while a large (90-kilogram) ground sloth, called *Megalocnus*, survived on the islands of Cuba and Hispaniola until 4,300 years ago. All sloth extinctions occurred shortly after the appearance of humans.

Despite the close correlation between the onset of human hunting and extinctions, there has been much debate about whether hunting alone—Martin's *Blitzkrieg* model—is responsible for the American megafauna's demise. Anti-*Blitzkrieg* proponents have pointed out that there have been very few finds of Clovis spears embedded in megafauna skeletons (the 'smoking gun' evidence of the Palaeolithic), and have argued that the earliest people are unlikely to have been sufficiently abundant to kill millions of animals and cause extinction. However, modelling has shown that for large, slow-to-reproduce animals, it is only necessary to kill a small percentage of the population for a species to decline rapidly.

Island extinctions in the Holocene

By 10,000 years ago, modern humans had colonized all the major continents, with the inevitable exception of Antarctica, but there were still plenty of islands to discover and new species to kill. In all cases the arrival of humans in the Holocene was followed by a wave of extinctions of the endemic species. There is general consensus that anthropogenic habitat destruction and overkill were responsible for all these extinctions; climate change does not figure in Holocene extinction models.

The large island of Madagascar saw the elimination of many species between 800 and 1,000 years ago. The losses included the elephant bird (*Aepyornis*) that weighed up to 500 kilograms, pygmy hippos, and several species of giant lemurs. Evidence for human colonization on Madagascar dates back 2,200 years and so there was appreciable overlap between people and the megafauna (although there was a similar lag time between human arrival and Pleistocene megafaunal extinctions seen, for example, in Australia and South America). The *Sporormiella* record from the region shows the extinctions coincided with the abrupt disappearance of large herbivore populations (and their dung) in less than 200 years, indicating that Madagascan wildlife population sizes had been stable until a threshold, perhaps caused by increased hunting pressure or accelerated forest clearance, had been crossed.

Other island extinction events occurred much sooner after first contact with humans. The colonization of New Zealand saw one of the swiftest destructions of native fauna ever recorded. Polynesian settlers first landed in 1280 where they found islands teeming with endemic animals, including nine species of giant flightless birds called moa. By around 1430, all the moa had gone, along with Haast's eagle, the largest raptor ever recorded. Weighing up to 16 kilograms, this large predator's diet had probably centred on the moa. The discovery of the island of Mauritius in 1598 by the

Dutch navy represents one of the last encounters between humans and an untouched island fauna. The result was perhaps inevitable; by 1662 the native dodo was extinct.

We have considered these late Pleistocene and Holocene extinctions of megafauna (and some not-so-'mega' fauna) region by region, but the debates about their causes (humans versus climate change) are very similar. Those who argue against human-driven extinction highlight the paucity of direct archaeological evidence for hunting. For example, very few North American megafauna sites are associated with Clovis spear points, and somewhat ironically, the best examples involve the butchery of bison which is one of the few megafaunal species to survive to the present day. Many of the extinctions also postdate the arrival of humans, especially in Eurasia where the mismatch is >20,000 years, but the earliest tentative records of human colonization of the Americas and Australia are also several thousand years before extinction. Conversely, it should be noted that no megafauna went extinct before humans arrived.

Those who prefer climate-driven extinction arguments have a major problem to overcome because the late Pleistocene climate changes were no different from the earlier extinction-free intervals of the Ice Age. Whatever caused it, the intense cooling of the Younger Dryas episode was the only exceptional climatic event, but only the North American megafauna and a few Eurasian species went extinct around this time. The selective loss of only large animals (and those with low reproductive rates) is also not well-explained by climate change models. Under the normal 'rules' of extinction, highest losses generally occur among species with a relatively limited habitat range, but the Pleistocene extinctions were fundamentally different. Many of the megafaunal species inhabited a vast geographic extent: the woolly mammoth and woolly rhino ranged across the whole of Eurasia and North America. Climate-driven extinction models invoke habitat change, such as the loss of tundra to advancing forests, to explain mammoth extinctions. These arguments do not account for the continuous presence of extensive tracts of all Pleistocene habitats up to the present day. In contrast, the extinctions can simply be ascribed to the observation that humans tend to hunt easy to find big animals.

Although scientific study should be free of such influences, there is also an underlying sociopolitical agenda to the Pleistocene megafauna debates. Early human cultures are often assumed to have existed in harmony with their environment in counterpoint to the destruction being wrought today. Similarly, indigenous peoples, with their belief in the perceived values of their ancestors, may take umbrage if their forefathers are accused of the wholesale destruction of ecosystems, even when the evidence can be overwhelming. And ardent proponents of the dangers of future climate change inevitably favour climate-driven models for past extinctions. Thus, Stephen Wroe and colleagues from the University of Sydney wrote in 2006 that 'by minimizing the role of climatic influences on Pleistocene events [and instead blaming humans], we risk overlooking salient lessons from the past in a world now facing significant climate change'. Alternatively, it could be argued that, by being too quick to blame climate change, we fail to recognize the ability of humans to eradicate entire terrestrial ecosystems all on their own. The *Sporormiella* data are especially important because they suggest that megafauna populations generally remain stable until an irreversible tipping point is reached, whereupon waves of extinction rapidly follow within a few hundred years—an observation of potential significance for the remaining megafauna in Africa.

Neanderthals and human extinctions

The genus *Homo* has been immensely successful over the past two million years. From its African heartland, at least a dozen species have evolved, and several have spread far and wide. However, from the viewpoint of diversity, *Homo* is currently in serious trouble, because we are the only species left. Studies on human evolution rarely focus on this aspect of our family tree, with its numerous dead-end branches, but instead they are typically concerned with ancestral relationships and the step-by-step acquisition of the attributes, such as bipedality and a large brain, that makes us human.

The problem with investigating *Homo* extinctions is the sheer rarity of fossils. For example, *Homo rudolfensis*, one of the earliest species, is only known from a single skull, meaning that its first appearance in the fossil record was also its last. Obviously, it is highly unlikely that this species had such a short lifespan, but in the absence of any further evidence we cannot know. Neither is it possible to determine a species' longevity based on how primitive or advanced it is. For example, one of the most primitive human species, *Homo naledi*, found in Rising Star Cave in South Africa, shows many attributes of our cousins the australopithecines and was thought to be one of the first species to branch away from this group around two million years ago. This much is probably true, but when the first accurate dates from Rising Star Cave were obtained in 2017, to everyone's surprise the bones turned out to be only 280,000 years old. Therefore, it seems likely that *Homo naledi* survived for at least 1.7 million years and was one of the most successful of all human species (based on the criterion of longevity). After such a long lifespan why did *Homo naledi* go extinct? This is again an impossible question to answer because its fossils just turn up at a single point in time.

The only extinct human species for which the data are abundant enough to consider the cause of its extinction is one of the most recent species to disappear: *Homo neanderthalensis*. This has fostered a dynamic debate going back more than a century in which a panoply of diverse and sometimes bizarre extinction causes has been proposed, including pathogens, climate change, interbreeding, lack of appropriate clothing, and competitive exclusion by humans.

Most *Homo* species originated in Africa but the Neanderthals uniquely evolved in Europe around 250,000 years ago probably from *Homo heidelbergensis*, which had arrived from Africa around 300,000 years earlier. The *H. heidelbergensis* that stayed behind in Africa also gave rise to a new species, called *Homo sapiens*, around the same time, or a little earlier, but these near-cousins did not meet the Neanderthals for nearly 200,000 years. Compared to ourselves, *H. neanderthalensis* was much more powerfully built, slightly shorter, and possessed a slightly larger brain. There are also clear differences in the skull (Figure 19). Neanderthals had a weaker chin, strong eyebrow ridges, a low forehead, and a distinctly extended rear portion of the skull. Their lifespans appear to have been shorter, reaching maturity around 12 years of age and rarely living beyond 40. Individuals with crippling injuries are known to have lived into old age, suggesting that invalids were cared for, probably in social groups; and, from 100,000 years ago, they started to bury their dead (a behaviour otherwise unique to our own species). Studies of their teeth chemistry shows that their diet was predominantly meat, probably from large game like mammoths. To help them kill their prey Neanderthals developed an array of simple stone tools, like arrow

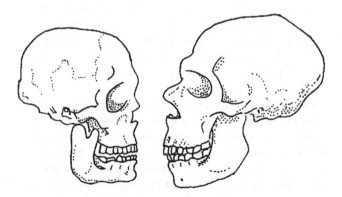

19. Comparison of the skulls of modern humans (left) and Neanderthals (right).

points and blades, which constitute part of the Mousterian Industry attributed to Neanderthals.

Early in their history the Neanderthals were content to remain in Europe, probably living in small hunter-gatherer groups; but from 100,000 years ago they started to spread eastwards and by 45,000 years ago their range spanned 8,000 kilometres from the Atlantic coast of Europe to deep into Siberia and the Middle East. The Neanderthals were clearly doing well at this stage in their history and they might have been predicted to enjoy a future of continued expansion, but at this point they ran into modern humans. The initial contact is seen in the Levant region, such as modern Israel, where human and Neanderthal remains and their tools are found interlayered in sediments from around 60,000 to 48,000 years ago. These alternations have been attributed to climatic oscillations, with Neanderthals occurring during inferred cooler intervals and modern humans during the warmer times. This delicate balance of the Neanderthal/human front line changed dramatically around 45,000 years ago with the rapid expansion of modern humans out of Africa into the Middle East and then Europe and Asia. It was accompanied by an equally rapid retreat of the Neanderthals, and by 40,000 years ago they were extinct.

Early ideas for the extinction of Neanderthals included the notion that they interbred with the much more abundant humans and so became 'diluted' out of existence. The sequencing of the Neanderthal genome in 2010 debunked this idea but it did show that most modern humans possess a small percentage of Neanderthal genes, indicating limited interbreeding, probably during their first encounters in the Levant. The exceptions are those people from sub-Saharan Africa whose ancestors never met the Neanderthals and so do not have any of their genes. That modern humans and Neanderthals could mate with each other and produce viable offspring shows how closely related the two species were. However, after a potentially convivial first encounter, subsequent human expansion seems to have been to the detriment of the Neanderthals. How much temporal overlap there was between the two *Homo* species during this invasion has been much discussed.

The arrival of modern humans in Europe is seen in a dramatic change in archaeological artefacts that defines the boundary between the Mid and Late Palaeolithic eras. Thus, the simple stone tools of the Mousterian Industry rapidly give way to a much more diverse array of artefacts produced by modern humans. These new Aurignacian artefacts included arrows, harpoons, fishhooks, needles, numerous types of knife, and a diverse array of jewellery (made from ivory, bone, and shells) and carved figures. Cave painting in Europe also began around 40,000 years ago, and shows the advanced artistic capabilities we expect of modern humans. There was, however, a short-lived and intriguing transitional industry in southern France and northern Spain, named the Châtelperronian. This is distinguished by bladed stone tools and ivory ornaments which are more sophisticated than Mousterian artefacts but significantly less impressive than Aurignacian ones. This raises the question of whether the Neanderthals improved their technology following cultural interactions with humans in the final few thousand years of their existence. The presence of Neanderthal bone fragments at some Châtelperronian sites has been used to link the makers with tools, but of course such evidence could also simply record the presence of butchered Neanderthals at sites occupied by humans.

The last Neanderthals may have lived in Spain and Portugal where, some have suggested, their population density was sufficiently high to hold back the influx of *Homo sapiens*. Specifically, it has been argued that the River Ebro in northern Spain was a front line that was only breached around 37,000 years ago. This was followed by the rapid extirpation and extinction of the remaining Neanderthal population. Unfortunately, problems with dating evidence from around 40,000 years ago makes it difficult to verify this story. The issue lies with the widely applied technique of carbon dating which uses the decay of carbon-14 as a radioactive clock. This isotope has a half-life of 5,730 years and so samples from >35,000 years ago retain very little of their original carbon-14, making them highly susceptible to contamination by trace amounts of modern carbon. Thus, artefacts and bones from ~40,000 years ago often yield dates that are a little younger than they should be, and this has probably happened in the case of the last Neanderthal remains from Spain. The latest, improved dating efforts suggest that the European Neanderthals disappeared very rapidly across the entire continent around 40,000 years ago, indicating there was no refuge or final stronghold in Spain.

Currently, the most favoured but nonetheless debated causes for Neanderthal extinction are competitive exclusion and climate change. The latter model focuses on Heinrich Event 4, one of six such episodes recorded in the past 60,000 years. These events consist of short-lived (~1,000-year) intervals when icebergs became abundant in the North Atlantic. They seem to be associated with changed rainfall patterns and, for some examples but not all, cooling of terrestrial temperatures. The origin of this climatic phenomena is poorly understood and need not concern us here, but Event 4 coincided with the Neanderthal extinction and so has been directly linked to their demise. It is argued that conditions became cooler (which may have contributed to an already ongoing cooling trend) causing Neanderthal populations to fragment and become extinction-prone. Adding detail to this scene, a recent idea suggests that the Neanderthals were incapable of making warm clothing like *Homo sapiens* (with their needles and sewing ability) and so were less able to adapt to climate cooling. Clearly though, there are many problems with this entire scenario. Neanderthals evolved in the cool climate of Europe, in contrast to the modern humans who had recently arrived from the hot African savanna, and so the former had already experienced much colder conditions in their history. If any species was going to suffer during a cold snap, surely it would be the newly arrived African immigrants. And why were Neanderthals the only species affected by Heinrich Event 4, which was just one of several such events and not even the most intense one?

Competitive exclusion models suggest that human hunting pressure on the European megafauna deprived the Neanderthals of their prey. However, the decline in the megafauna occurred much later—mammoths were still common throughout the region until 15,000 years ago. Other scenarios suggest, rather vaguely, that the Neanderthals were unable to adapt fast enough to human-altered environments. More likely, but rarely explicitly stated in the archaeological literature, humans may have simply hunted down and killed the Neanderthals. Despite being individually strong, the Neanderthals had to face an intelligent and cunning species with sophisticated weaponry that was likely to have considerably outnumbered them. We know only too well, from numerous examples in more recent history, the destructive capabilities of our species, especially when invading new territory.

As with all scientific debates, there are those who try to reconcile arguments with compromise: many suggest that the Neanderthal's extinction was caused by a combination of climate change and human-driven habitat changes in the same way that the dinosaur extinction is sometimes ascribed to a combination of *both* meteorite impact and large-scale volcanism. None of these attempts to accommodate both viewpoints are very satisfactory because they raise unanswerable questions: which was more important, climate change or human factors? If *Homo sapiens* had failed to appear in Europe would the Neanderthals still be with us? If there had been no Chicxulub impact would the Deccan eruptions have still caused mass extinction? In fact, a look back at the history of extinction research over the past decades shows that many debates have been answered without a need to fudge the issue by blaming everything at once.

The K-Pg impact theory was first proposed in 1980 and ten years later, Niles Eldredge, one of the great palaeontologists of his generation, published a valuable summary of the ongoing debates as they then stood. Reading Eldredge's *The Miner's Canary: Unravelling the Mysteries of Extinction* thirty years after it was written shows that many of the ideas from that time are either no longer current or have been shown to be wrong. Popular causes of extinctions, such as ice ages and major sea-level fall, are no longer considered tenable (with exceptions like the Late Ordovician mass extinction) because new evidence, especially on the timing of extinctions, shows no link. Other causes such as ozone depletion and ocean acidification had barely entered the lexicon of debate in 1990 but are now viable kill mechanisms. Eldredge even throws in an optimistic note that human-effected global warming may help to offset an impending glaciation—how our worries have changed!

One aspect of Eldredge's book that has not changed since 1990 was his serious concern for the impending extinction of much of the world's remaining fauna due to habitat destruction and over-exploitation driven by population growth. This issue has not gone away and it is sobering to think that, since Eldredge wrote his words, the world has become even more crowded: humanity has increased by another 1.2 billion and stands at 7.2 billion today (with the possibility of reaching 9.6 billion by 2050). The impending sixth mass extinction is a reality that requires much urgent action. Recovery from mass extinctions takes millions of years, which is forever on a human timescale. Add current worries about climate change, which may not happen next year but will almost certainly happen in the next few centuries, and it is clear that extinction is going to remain a global issue for the foreseeable future.

Further reading

Chapter 1: Why extinctions happen

Bellard, C., Bertelsmeier, C., Leadley, P., Thuiller, W., and Courchamp, F. 2012. Impacts of climate change on the future of biodiversity. *Ecology Letters*, **15**, 365–77.

Benitez-Lopez, A., Alkemade, R., Schipper, A.M., et al. 2017. The impact of hunting on tropical mammal and bird populations. *Science*, **356**, 180–3.

He, F.L. and Hubbell, S.P. 2011. Species–area relationships always overestimate extinction rates from habitat loss. *Nature*, **473**, 368–72.

Simberloff, D. 1985. The proximate causes of extinction. In: *Patterns and Processes in the History of Life*, D.M. Raup and D. Jablonski (eds), pp. 259–76. Springer-Verlag, Berlin.

Wiens, D. and Slaton, M.R. 2012. The mechanism of background extinction. *Biological Journal of the Linnean Society*, **105**, 255–68.

Chapter 2: Extinction today and efforts to stop it

Barnovsky, A.D., et al. 2011. Has the Earth's sixth mass extinction already arrived? *Nature*, **471**, 51–7.

Ceballos, G., Ehrlich, P.R., Barnovsky, A.D., Garcia, A., Pringle, R.M., and Palmer, T.M. 2015. Accelerated modern human-induced species losses: Entering the sixth mass extinction. *Science Advances*, 1, e1400253.

Girling, R. 2014. *The Hunt for the Golden Mole: All Creatures Great and Small and Why They Matter*. Vintage Books, London.

Quammen, D. 1997. *The Song of the Dodo: Island Biogeography in an Age of Extinctions*. Scribner, New York.

Williams, G. 2011. 100 *Alien Invaders: Animals and Plants that are Changing our World*. Bradt Travel Guides Ltd, London.

Chapter 3: Extinction in the past

De Vos, J.M., Joppa, L.N., Gittleman, J.L., Stephens, P.R., and Pimm, S.L. 2014. Estimating the normal background rate of species extinction. *Conservation Biology*, **29**, 452–62.

Jablonski, D. 2008. Extinction and the spatial dynamics of biodiversity. *Proceedings of the National Academy of Sciences*, **105**, 11528–35.

Raup, D.M. 1994. The role of extinction in evolution. *Proceedings of the National Academy of Sciences*, **91**, 6758–63.

Chapter 4: The great catastrophes

Alvarez, L.W., Alvarez, W., Asaro, F., and Michel, H.V. 1980. Extraterrestrial cause for the Cretaceous-Tertiary extinction. *Science*, **208**, 1095–108.

Erwin, D.H. 2006. *Extinction: How Life on Earth Nearly Ended 250 Million Years Ago*. Princeton University Press, Princeton, NJ.

Hallam, A. and Wignall, P.B. 1997. *Mass Extinctions and Their Aftermath*. Oxford University Press, Oxford.

MacLeod, N. 2013. *The Great Extinctions: What Causes Them and How They Shape Life*. Natural History Museum, London.

McGhee Jr, G.R. 2018. *Carboniferous Giants and Mass Extinction: The Late Paleozoic Ice Age World*. Columbia University Press, New York.

Wignall, P.B. 2015. *The Worst of Times: How Life on Earth Survived Eighty Million Years of Extinctions*. Princeton University Press, Princeton, NJ.

Chapter 5: How to kill nearly everything

Courtillot, V. 1999. *Evolutionary Catastrophes: The Science of Mass Extinction*. Cambridge University Press, Cambridge.

Hallam, A. 2004. *Catastrophes and Lesser Calamities: The Causes of Mass Extinctions*. Oxford University Press, Oxford.

Raup, D.M. 1991. *Extinction: Bad Genes or Bad Luck?* W.W. Norton & Co., New York.

Wolbach. W.S., et al. 2018. Extraordinary biomass-burning episode and impact winter triggered by the Younger Dryas cosmic impact ~12,800 years ago: Ice cores and glaciers. *Journal of Geology*, **126**, 165–84.

Chapter 6: What happened to the Ice Age megafauna?

MacPhee, R.D.E. 2019. *End of the Megafauna: The Fate of the World's Hugest, Fiercest, and Strangest Animals*. W.W. Norton and Co., New York.

Sarmiento, E., Sawyer, G.J., and Milner, R. 2007. *The Last Human: A Guide to Twenty-Two Species of Extinct Human*. Yale University Press, New Haven.

Stuart, A.J. 2015. Late Quaternary megafaunal extinctions on the continents: A short review. *Geological Journal*, **50**, 338–63.

Turvey, S.T. ed. 2009. *Holocene Extinctions*. Oxford University Press, Oxford.

Index

SOCIAL MEDIA
Very Short Introduction

Join our community

www.oup.com/vsi

- Join us online at the official Very Short Introductions **Facebook** page.
- Access the thoughts and musings of our authors with our online **blog**.
- Sign up for our monthly **e-newsletter** to receive information on all new titles publishing that month.
- Browse the full range of Very Short Introductions online.
- Read **extracts** from the Introductions for free.
- If you are a teacher or lecturer you can order inspection copies quickly and simply via our website.

CHAOS
A Very Short Introduction
Leonard Smith

Our growing understanding of Chaos Theory is having
fascinating applications in the real world - from technology to
global warming, politics, human behaviour, and even gambling
on the stock market. Leonard Smith shows that we all have an
intuitive understanding of chaotic systems. He uses accessible
maths and physics (replacing complex equations with simple
examples like pendulums, railway lines, and tossing coins) to
explain the theory, and points to numerous examples in
philosophy and literature (Edgar Allen Poe, Chang-Tzu, Arthur
Conan Doyle) that illuminate the problems. The beauty of fractal
patterns and their relation to chaos, as well as the history of
chaos, and its uses in the real world and implications for the
philosophy of science are all discussed in this *Very Short
Introduction*.

> '...Chaos...will give you the clearest (but not too painful idea) of
> the maths involved... There's a lot packed into this little book, and
> for such a technical exploration it's surprisingly readable and
> enjoyable - I really wanted to keep turning the pages. Smith also
> has some excellent words of wisdom about common
> misunderstandings of chaos theory...'

popularscience.co.uk

www.oup.com/vsi